SpringerBriefs in Earth Sciences

More information about this series at http://www.springer.com/series/8897

Emily So

Estimating Fatality Rates for Earthquake Loss Models

Emily So
Department of Architecture
University of Cambridge
Cambridge
UK

ISSN 2191-5369 ISSN 2191-5377 (electronic)
SpringerBriefs in Earth Sciences
ISBN 978-3-319-26837-8 ISBN 978-3-319-26838-5 (eBook)
DOI 10.1007/978-3-319-26838-5

Library of Congress Control Number: 2015957038

Printed on acid-free paper

This Springer imprint is published by SpringerNature
The registered company is Springer International Publishing AG Switzerland

Preface

Earthquakes themselves do not kill. It is the vulnerability of the natural and built environment that does, be it through landslides, tsunamis, poorly built infrastructure or housing. Nearly 700,000 people have died in earthquakes during this century's first decade and over 80 % of these have died from collapses of poor housing. The cruel counterpoint to this is that engineers know how to build in earthquake-prone areas. The 16th September 2015 M8.3 Coquimbo earthquake in Chile is a testament to what an earthquake resilient community can do. The low death count is entirely attributed to earthquake preparedness through education and building code enforcement in Chile. Most of the deaths of the last few decades are preventable ones. As a scientific community, we have a duty of care to people living in earthquake-prone countries. Given the physical, social and economic constraints of the most vulnerable communities around the world, the challenges and opportunities for scientists and engineers lie in communicating earthquake risks, based on the rigorous scientific evaluation of earthquake hazard to decisions makers and affected communities to provide life safety through education and construction.

As a first step, we must assess both potential physical and social losses that can result from a future event to raise awareness to the threat and also help decision makers prioritise. This book brings together empirical evidence of the fatalities from 25 earthquakes over 40 years and proposes a set of judgment-based fatality rates for use in earthquake loss estimation models. The main aims of this research are to improve the current understanding of fatalities from earthquakes and to encourage researchers in the field pay greater attention to the collection and evaluation of casualty data in the future.

Acknowledgments

Thanks go to the US Geological Survey and its PAGER team for their support over the author's Mendenhall fellowship (2010–2011) during which time the main elements of this research were accomplished and to Antonios Pomonis who contributed to this book and was party to the derivation of the judgment-based fatality rates in Table 5.1. He has been working in the field of casualty estimation for over 20 years and had carried out extensive studies in the past on fatalities from earthquakes. Lastly, thanks also go to Professor Robin Spence and Hannah Baker for their input and support in getting the manuscript ready for publication.

Contents

Abbreviations

EEFIT	Earthquake Engineering Field Investigations Team, UK
EERI	Earthquake Engineering Research Institute, USA
FDMA	Fire and Disaster Management Agency, Japan
GEM	Global Earthquake Model
MRF	Moment Resisting Frame
RC	Reinforced Concrete
URM	Unreinforced Masonry
USGS	United States Geological Survey, USA

Chapter 1
Introduction

At present, there are no globally applicable fatality rates used in loss estimation models and within the research of earthquake casualties, there is certainly a lack of comprehensive data collection in the field following actual events. As the population living in areas exposed to seismic activity continues to increase and settlements are located in more hazardous areas, it is vital that engineers understand the implications of these social changes and use tools available to them to inform decision makers and response teams to mitigate and react to situations of mass earthquake casualties. In Marano et al.'s (2009) assessment of earthquake deaths in the period September 1968 to June 2008, they concluded that 77.7 % of the total earthquake deaths were due to ground shaking-related building damage; if we consider the official death toll of 316,000 for the Haiti earthquake of 2010 (though this figure is much disputed), this would increase the contribution from building damage to over 80 %. The main cause of deaths and injuries, as suggested by previous studies (Coburn and Spence 2002; Marano et al. 2009) are due to damage and especially due to collapse of buildings. Therefore only in understanding the associations between the levels of damage and the severity and types of injuries can one make headway in deriving realistic estimates of human casualties in future earthquakes.

As a discipline, engineers and architects have been able to influence changes in building codes and control in many parts of the developed world (Spence 2007) but have struggled to achieve a significant impact on most developing countries, where the pace of urbanisation during recent decades has been tremendous. In his Mallet Milne lecture of 2007, Spence carried out a questionnaire survey amongst practising engineers from around the world. Of those countries categorised in the 'growing risks' category (countries such as Nepal and Iran), the replies would suggest that in some cities, nearly 90 % of the building stock may be classified as seismically unsafe, as shown in Fig. 1.1.

However, as Bilham noted in his Mallet Milne lecture of 2009, the reasons for this large percentage of seismically unsafe buildings, in addition to poverty, are attributed to indifference, ignorance and corrupt practices, not due to an absence of engineering knowledge or competence. He continued to say that "never has a generation of earthquake engineers been faced with such a grave responsibility to exercise their skills, both political and technical, as now" (Bilham 2009). In this

E. So, *Estimating Fatality Rates for Earthquake Loss Models*,
SpringerBriefs in Earth Sciences, DOI 10.1007/978-3-319-26838-5_1

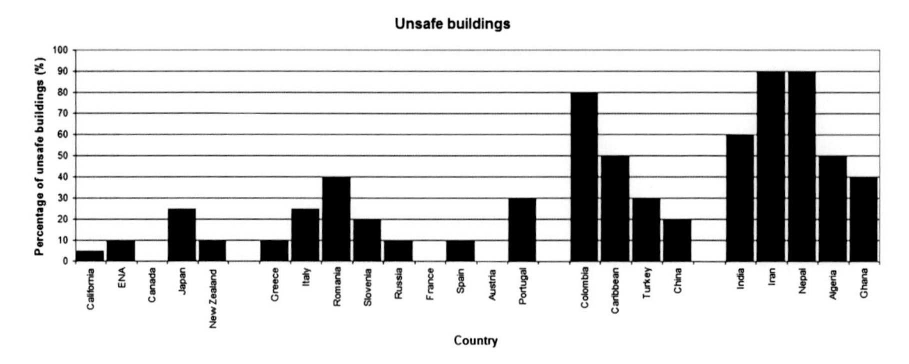

Fig. 1.1 Graph summarising responses from international earthquake engineers on percentage of seismically unsafe buildings in their country. (*Source* Spence 2007)

century, earthquakes that had little impact on the pre-existing villages and towns in a region will be shaking urban agglomerations housing upwards of 12 million people. Figure 1.2 shows the locations of the world's megacities, housing more than two million people and their proximity to zones of the high plate boundary strain rate, as calculated by Kreemer et al. (2003). Areas of high strain (in brown) are where recurrence intervals of large earthquakes are of the order of 100 to 200 years.

Supporting Bilham's statement of the challenges faced by society and engineers, evidence over the past 112 years since 1900 confirms an increase in the actual

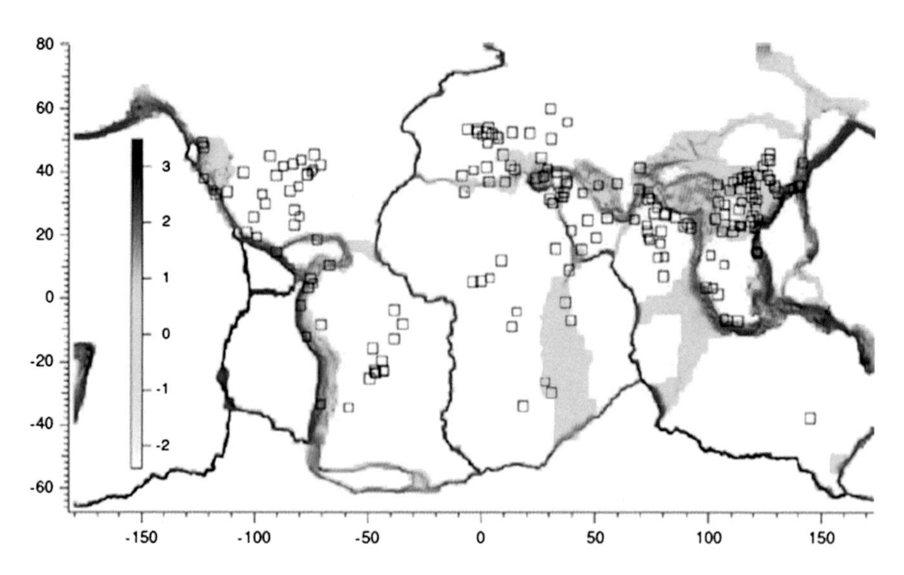

Fig. 1.2 The location of 194 supercities (each with a 2005 population exceeding 2 million) and their proximity to zones of high plate boundary strain rate (shaded from Kreemer et al. 2003). (*Source* Bilham 2009)

number of reported fatalities from earthquakes (1.05 million in 1900–1955 vs. 1.48 million in 1956–2011), although because uncertainties about the actual death toll in some of the most fatal earthquakes since 1900 are considerable (e.g., 2010 Haiti, 1999 Kocaeli, 1976 Tangshan, 1948 Ashqabad, etc.) such simple comparisons are rather tenuous. However, a temporal analysis of the fatal earthquakes since 1900 shows that there is an increase in the frequency of events with large death tolls when we compare the frequencies in the period 1900–1985 versus the last 30 years. This is especially true for earthquakes with very high death tolls. Table 1.1 shows earthquakes with over 10,000 deaths have become more frequent (increased by 43 %) over the last 30 years compared to 1900–1985, i.e., they occurred once per 1,096 days instead of once per 1,289 days in the pre-1985 period.

One should also note that in the last 30 years the contribution of the most deadly events (5000+ deaths) has grown to 92 % in terms of total deaths (from 86 % in 1900–1985), and even more for the mega-death events (10,000+ deaths, 87 from 80 %), as these have become more frequent.

In terms of estimating human casualties, current loss modeling practices, as highlighted in the benchmark study carried out by Spence et al. (2008) fall far short of what is required to reflect what happens in reality. Hampered by a lack of regional historic data, modelers are often left with what is available in models such as HAZUS (NIBS-FEMA 2006) in the United States to use in other countries where building inventory and their vulnerability attributes can be very different. Moreover, fundamental justifications for the commonly used expert-opinion casualty rates are often lost within models where the derivation of these rates is not explicitly explained. For example, how have the casualty rates in HAZUS been derived? There is no supporting documentation on the means of the derivation of these expert opinion casualty rates. Therefore, one of the central aims of this study is to dissect past earthquake fatality numbers, providing the evidence behind the fatality rates presented, making these judgment-based values accountable and realistic. The only way to do this is by exploring empirical casualty data from historical earthquakes.

Examining the reported fatalities in buildings based on their structure types, the way they fail and the vulnerability of occupants in the structures when they do collapse will inform modelers of the probability of death in a building. The assessment of lethality potential of different collapsed building types in this book is based on a systematic review of empirical casualty data obtained from past events and engineering judgment of likely collapse mechanisms when fatality rates cannot be inferred directly.

This manuscript has been divided into the following chapters. Chapter 2 reviews the methodology for the assessment of fatality rate assignments and provides a critique highlighting some important assumptions made. In addition, some accompanying information on possible causes of deaths, in particular literature on building collapses, are presented. The definitions used in the model are explicitly described in this chapter. Chapter 3 presents a review of the quality and quantity of supporting literature on fatalities from earthquakes and lists out the earthquakes examined in this study.

Table 1.1 Table shows the frequencies of high fatality earthquakes in the last century

Extent of Life Loss Class	No. of events (1900 to Aug' 85)	Total deaths	Percentage of total deaths (%)	Frequency in days per event	No. of events in the last 30 years (Sept' 85–Aug' 15)	Total deaths in the last 30 years (Sept' 85–Aug' 15)	Percentage of total deaths (Sept' 85–Aug' 15) (%)	Frequency (Sept' 85–Aug' 15)—in days per event
≥1	1,063	1,478,775	100.0	29	667	807,446	100.0	16
≥1,000	103	1,414,648	95.7	300	34	782,947	97.0	322
≥5,000	39	1,275,664	86.3	793	15	741,252	91.8	730
≥10,000	24	1,177,970	79.7	1,289	10	701,824	86.9	1,096
Total	1063	1,478,775			667	807,446		
Deaths/day		48				76		

The database runs from 1900 to 2015, containing 1,730 events and 2,286,221 deaths (Pomonis, written commun., 2015)

The main chapter of the book, Chap. 4 contains a systematic evaluation of evidence from recent events on the lethality potential of collapsed building of certain typologies. The main hypothesis for this model is that most ground shaking induced deaths are a consequence of building collapses. In this chapter, the differences between observed fatalities from recent events and notable variations in behavior among the same building typologies in different countries are also reviewed. Finally, a set of fatality rates for use in earthquake loss estimation models based on judgment from gathered empirical evidence are presented in the concluding Chap. 5.

References

Bilham R (2009) The seismic future of cities. Bull Earthq Eng 7(4):839–887

Coburn AW, Spence RJS (2002) Earthquake protection, 2nd edn. Wiley, Chichester, p 436

Kreemer C, Holt WE, Haines AJ (2003) An integrated global model of present-day plate motions and plate boundary deformation. Geophys J Int 154(1):8–34

Marano KD, Wald DJ, Allen TI (2009) Global earthquake casualties due to secondary effects: a quantitative analysis for improving rapid loss analyses. Nat Hazards 49. 10.1007/s11069-009-9372-5

NIBS-FEMA (National Institute of Building Sciences-Federal Emergency Management Agency) (2006) Multi-hazard loss estimation methodology, HAZUS-MH MR2 technical manual, prepared for the Federal Emergency Management Agency, Washington DC, United States, pp 727

Spence RJS (2007) Saving lives in earthquakes: successes and failures in seismic protection since 1960. Bull Earthq Eng 5(2):139–251

Spence R, So E, Jenny S, Castella H, Ewald M, Booth E (2008) The Global Earthquake Vulnerability Estimation System (GEVES): an approach for earthquake risk assessment for insurance applications. Bull Earthq Eng 6(3):463–483

Chapter 2
Main Assumptions of the Assignment Process

Most earthquake loss estimation models follow a modular process (e.g., US Geological Survey's (USGS) PAGER semi-empirical, LNEC-Loss, and SELENA), similar to the one described in Fig. 2.1 where estimated ground motion parameters, in this case taken from USGS ShakeMaps are fed into the population exposure and building inventories of the affected region. The probabilities of collapse are described by a set of vulnerability functions for particular building types for a specific location, and fatality estimates are then derived from the estimated proportion of collapses. The model will then produce loss estimates, whether in terms of numbers of building damage, human casualties or economic losses.

In the USGS PAGER system and other loss estimation models including FEMA's HAZUS (NIBS-FEMA 2006), it is assumed that most fatalities are caused by building collapses. This hypothesis is well supported by field observations and studies (Coburn and Spence 2002, Marano et al. 2009). It should be noted that fatalities from secondary hazards, such as fire-following earthquakes, landslides, mudslides, avalanches, rock falls and tsunamis could also be the main or an important cause of fatalities in an event, as evident in the Great Eastern Japan Earthquake (GEJE) and tsunami of March 2011. As most loss models are modular, it is possible in the future to include models for capturing fatalities due to secondary hazards, though this is not covered in this assessment.

2.1 Contributing Factors to Earthquake Fatalities

There are some key challenges to modeling earthquake fatalities as there are many contributing factors that could augment or diminish the final fatality number, some to having a significant impact. For example in the recent Nepal earthquake of April 2015, there were fortunately fewer fatalities than anticipated by the earthquake engineering committee with knowledge of the types of buildings in the affected region. The reasons behind the low death toll of the M7.8 Gorkha earthquake are certainly of interest to the earthquake engineering community. Seismologists who have worked extensively in the region like Professor Jean Philippe Avouac was able to determine with the GPS instruments in the area, that the geologic oscillations

E. So, *Estimating Fatality Rates for Earthquake Loss Models*,
SpringerBriefs in Earth Sciences, DOI 10.1007/978-3-319-26838-5_2

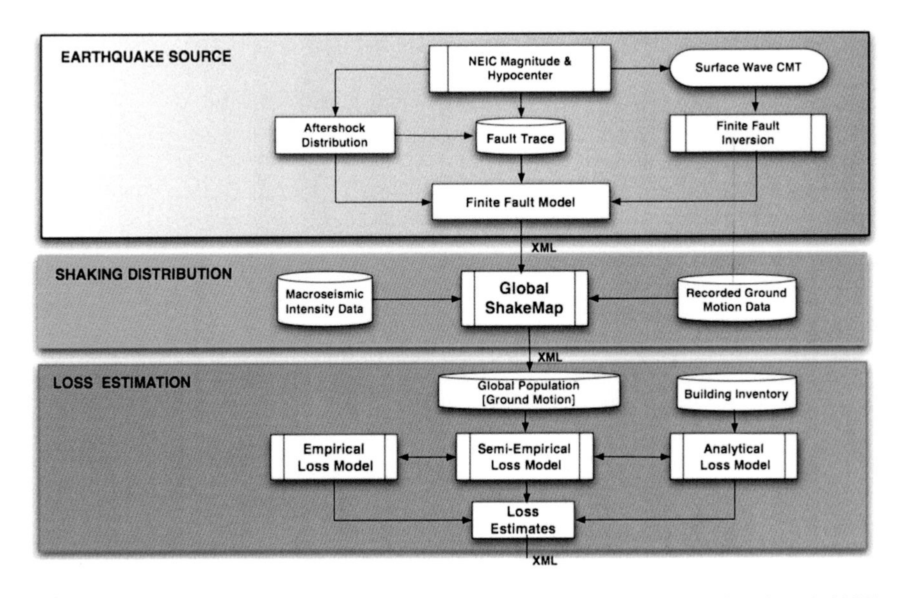

Fig. 2.1 The loss estimation process in the USGS PAGER system (*Source* Jaiswal et al. 2009)

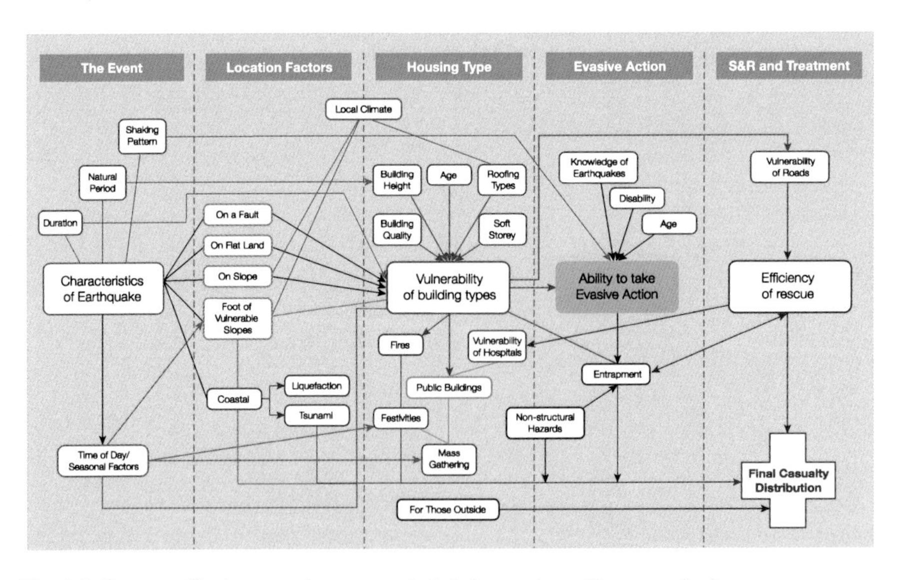

Fig. 2.2 Factors affecting casualty rates and their interactions (*Source* author)

would have damaged taller buildings and therefore left the shorter, more vulnerable ones intact (Galetzka et al. 2015). Therefore an argument would be that the actual physical damage was much less and the casualties were less than expected as a result. The characteristics of ground motions is therefore an important factor.

Figure 2.2 is taken from the author's dissertation exploring the factors contributing to fatalities from earthquake (So 2009). As shown, modelling casualties is a complicated procedure with many dependencies.

One important omission in the proposed method is that factors other than building collapse contributing to the final ground shaking-related death toll have not been included (e.g., fatalities that occur in buildings that did not collapse; outdoor fatalities due to a multitude of causes such as debris falls, accidents; collapse of non-building structures such as bridges, dams, and industrial installations; fatalities related to heart attacks or other previous medical conditions). The collapse of high-occupancy buildings such as schools, churches, theatres, etc., is a factor that can amplify life loss significantly. In addition, other contributing or differentiating factors to the probability of death occurrence due to ground shaking include the quality of the building stock (standard and non-standard) and its collapse probability, the availability of escape routes and egress percentages, the density of habitation, amongst others (So 2009). The main reason for not including these factors is that they are very difficult to quantify and therefore it is not always possible to separate these contributing factors out from the empirical data collected in the field. An additional challenge is related to the time of occurrence, and the temporal-seasonal patterns of building occupancy levels at any specific earthquake location, as well as the occurrence of foreshocks. Other factors, such as the general preparedness of the communities and their search and rescue capabilities, which in turn are affected by the extent of the damage to the buildings, infrastructure and by potential difficulties in accessing settlements due to blocked roads or remote location, also play an important role. It is important for users to acknowledge these factors and the impact they may have on the final death toll of an earthquake as illustrated from some of the recent earthquakes discussed in this chapter.

2.1.1 Accessibility to the Affected Areas

The inability to access the towns of Muzzafarabad in Azad Jammu Kashmir, Pakistan in 2005 and Ying Xiu in China in 2008 due to blocked roads by landslides and rockfalls and inclement weather conditions until days after the earthquake were considered to be significant contributors to the high town-level death tolls. In Ying Xiu, 60 % of the town's population was reportedly killed even though there were still intact buildings in this heavily damaged town (Pomonis written commun., 2011). It can be postulated that many died as a result of exposure and prolonged delay in rescue as roads and tunnels into Ying Xiu were extensively damaged as shown in Fig. 2.3.

Fig. 2.3 Aerial view of the devastation in Ying Xiu after the Wenchuan earthquake of 2008 and the collapsed Bai Hua bridge into Ying Xiu (*Source* Author, and Yong et al. 2008)

2.1.2 Lucky Escapes

On the other hand, the Pisco, Peru event in 2007 and the 2009 L'Aquila earthquakes had lower than expected death tolls when compared with the observed physical damage to the building stock. The mainshock of the Pisco earthquake was preceded by a weaker motion when people reacted and were able to get out before their weak adobe buildings collapsed on them (So 2009). In L'Aquila, the epicenter was further away from dense population centers and the event was smaller than Pisco, but two further factors were thought to contribute to the relatively low death toll there. Firstly, in the two months prior to the 6th April event, the area had been subjected to a series of small earthquakes. In the town of Paganica, (where approximately 90 % of the masonry houses were damaged but the official number of fatalities is seven), interviewed survivors said they had slept in their cars after feeling an earthquake tremor four hours before the main event (EEFIT 2009). Secondly, the area contains many weekend (second) homes for people living in Rome and other large Italian cities, and as the earthquake happened in the early hours of Monday morning, many residents had already left the region (EEFIT 2009) (Fig. 2.4).

In addition, other factors such as characteristics of the ground motion and site conditions contribute to the final death toll (So 2009). The aforementioned contributory factors are critical for some events in augmenting or significantly reducing the number of victims. There is anecdotal evidence of the influence of these factors but they are hard to quantify, therefore adding to the uncertainty of estimating fatalities. This reflects the reality of post-earthquake data collection and the added complexity of assessing casualty data.

Fig. 2.4 Damaged masonry houses in Paganica after the 2009 L'Aquila earthquake in Italy (*Source* Author)

2.1.3 Extreme Cases of Catastrophic Collapses

In most cases, a high fatality earthquake is influenced by the failure of a dominant but weak building type (e.g., Haiti 2010, Bam 2003, Kocaeli 1999, Neftegorsk 1995, Killari 1993, Armenia 1988 and Mexico City 1985) and the sheer number of collapses of these buildings add to the large death tolls as they would require tremendous search and rescue resources. In reviewing recent earthquake casualty data in detail, four villages and towns in particular, exhibited disproportionate death tolls to the damage experienced (see Table 3.1 for list of events). In Tabas, Iran, in the 1978 earthquake, 85 % of the town's inhabitants were killed and in Ghir, Iran, 67 % of the inhabitants were killed, after the earthquake of 1972. The extensive number of collapsed structures meant that there were very few able bodied and untrapped people to help in search and rescue, which could have led to the extreme death rates.

The Bam earthquake of 2003 in Iran is one such event where around 26,800 people were killed in the M6.6 earthquake, mostly by collapses of weak adobe blocks in mud mortar masonry and subsequent asphyxiation (28 % of Bam's population was killed in this earthquake) (Fig. 2.5).

However, there are some earthquakes where the death tolls are governed by few rare cases of catastrophic collapses. This was the case for El Asnam (Algeria) in 1980 where it is estimated that at least 1,000 people died, out of 6,500 for the entire event, due to the collapse of a single building (the Cite An Nasr Complex) and hundreds due to the collapse of the Grand Mosque (as the earthquake occurred on a Friday's prayer time). In 1983, up to a third of the deaths of the Popayan earthquake in Colombia were due to the collapse of the Cathedral; similarly, the collapse of four reinforced concrete buildings contributed to 366 deaths out of 1,500 in San Salvador city in 1986. A more recent example is the Cathedral de San Clemente in Pisco which killed over 140 people (there were only two survivors from the roof collapse) after the M8.0 event in Peru in 2007.

Fig. 2.5 The aftermath of the Bam earthquake in 2003. About 26,800 people were killed in the M6.6 earthquake in the ancient city of Bam, about 600 miles southeast of Tehran, Iran (Credit: Wikimedia, Photographer: Marty Bahamonde)

Most recently in the Christchurch New Zealand earthquake in 2011, the collapses of two mid-rise reinforced concrete buildings accounted for 75 % of the final death toll of 181 people. The figure below shows the collapse of one of the two buildings, the Pyne Gould Corporation (PGC) building (Fig. 2.6).

Fig. 2.6 Catastrophic collapse of the Pyne Gould Corporation (PGC) Building in the Central Business District of Christchurch, New Zealand (*Source* Flickr)

These case studies although difficult to incorporate in general loss models (the two buildings only accounted for 2 % of this building type in the Central Building District in Christchurch, New Zealand) can inform our understanding of the survivability of occupants in such buildings. The studies not only provide useful accounts for search and rescue research but also offer low probability and high consequence limits for loss estimation.

2.2 Definition of Collapse

The principle of the stated method is that the majority of fatalities are caused by building collapses induced by earthquake ground shaking. The definition of collapse is therefore crucial. Assessing damage to a building and what constitutes a collapse is subjective and has been the topic of debate amongst earthquake engineers for many decades. The definition is further complicated by the end users' needs. For example, an assessment carried out rapidly after an event to give an indication for temporary housing needs will yield very different results to a survey of a building's integrity for engineering research purposes. From a financial perspective, a higher collapse percentage is not necessarily worse than a damaged but structurally intact building. As recent examples in Christchurch (2011) showed, in some cases partial collapse can actually be more costly than total collapses due to the added cost of safely bringing down damaged structures in urban areas.

Using data collected with such loose definitions of collapse does pose problems. If the definition of complete collapse (D5) of "more than one wall collapsed or more than half of a roof dislodged or failure of structure members to allow fall of roof or slab" was used, taken from the Cambridge definition (Coburn et al. 1992), the actual volume reduction and therefore lethality potential would vary dramatically. For example, for weak masonry, 'collapsed' buildings can have volumetric reduction ranges from 10 to 100 % as shown in Fig. 2.7.

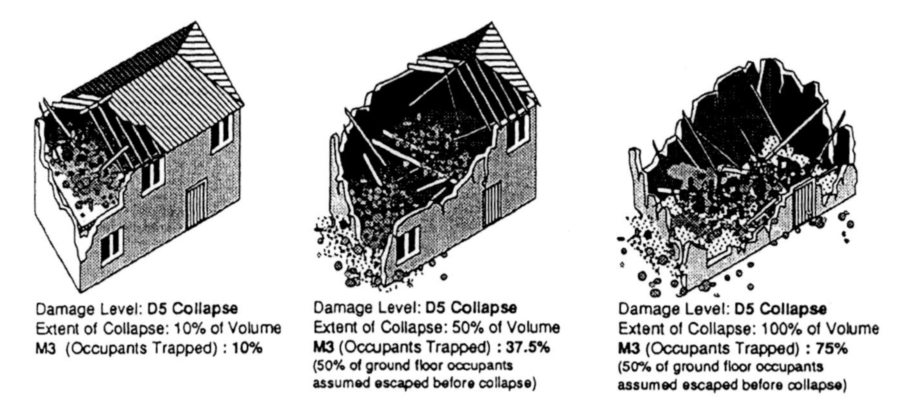

Fig. 2.7 Sketches showing the differences in volumetric reduction of a single collapsed building with implications on survivability of its occupants (*Source* Coburn et al. 1992)

The masonry collapses illustrated in Fig. 2.7 have an average volumetric reduction of 41 %. The survivability of occupants in buildings depends on its collapse mechanism and the volume loss to the structure as investigated in Okada's study (1996). Given this variation and its implications on casualty numbers and search and rescue (SAR) requirements, an assessment of possible collapse forms of buildings is necessary and presented in the next section.

2.3 An Assessment of Collapse Mechanisms

This section presents a preliminary attempt at capturing typical failure mechanisms of buildings within the damage category D5 (collapse) of damage assessments, in the context of fatality generation and therefore volumetric reduction. It is also important to understand failure modes for retrofitting buildings and mitigating future deaths.

Two main references were consulted in creating the suggested 'damage catalogue', including Schweier and Markus (2006) and Okada and Takai (2000).

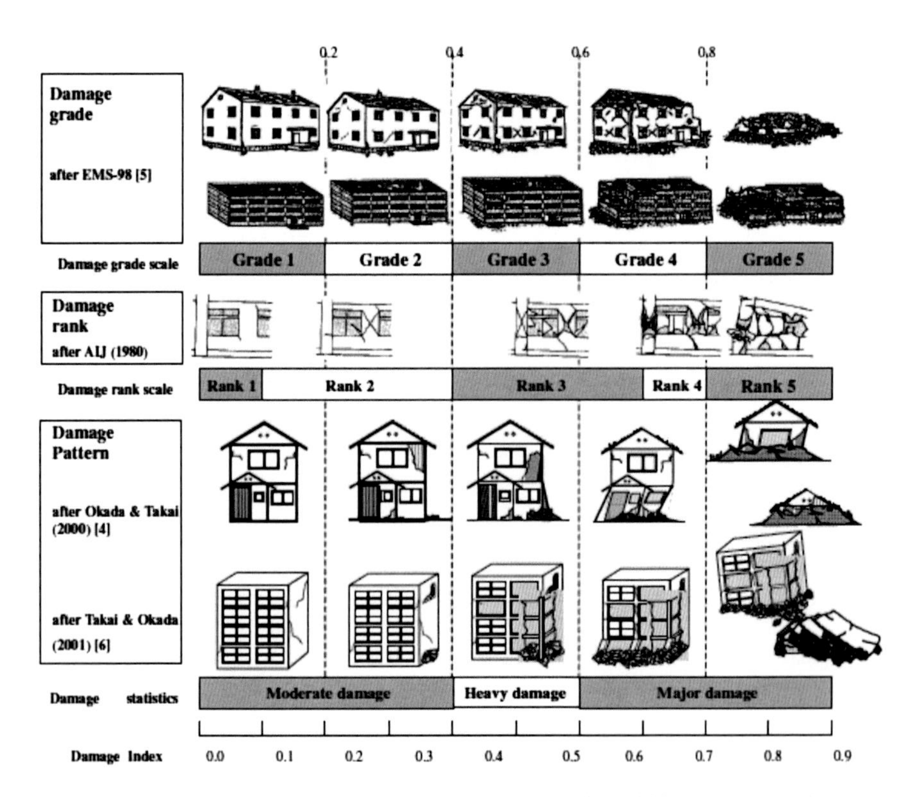

Fig. 2.8 Comparison of established damage scales (*Source* Okada 2004)

The first is a scheme developed by Schweier and Markus (2006) which includes a compilation of different damage types of reinforced concrete frame buildings typically occurring after earthquakes and their characterisation by geometrical features like volume reduction or inclination change. The Schweier and Markus categories are derived from various reconnaissance and damage reports as well as photographs of damaged buildings, which were collected and analysed in this study. However, since their catalogue was devised for damage recognition from remote sensing techniques, it contains an odd mixture of structural descriptions and satellite image observations. The second reference reviewed was the classification suggested by Okada and Takai (2000). Although concentrating on Japanese building types, Okada and Takai developed a damage assessment based on damage indices and following this work in Okada's paper in 2004, he compares the damage types from other sources as shown in Fig. 2.8 (Okada 2004).

Collapses related to what is termed damage indices 0.5–1.0 in Okada's scale have been examined and in the following section, concentrating on the contributions of the responsible structural components, we assess damage characteristics in the context of fatality generation.

The resulting damage catalogue is characterised in the following six groups:

1. Roof collapse
2. Single wall/vertical support failure
3. Multiple walls/vertical supports failure
4. Soft storey collapse
5. Complete failure of structural elements (pancake collapse)
6. Overturning/toppling.

An attempt to create a damage catalogue that encompasses all building types and possible failure mechanisms has been made. Each of these subtypes is explained in turn in the ensuring section.

2.3.1 *Roof Collapse (Failure of a Single Lateral Support)*

Roof collapse describes damage to more than half of a roof or failure of structural members to allow fall of a roof or slab, as shown in Fig. 2.9.

2.3.2 *Single Wall/Vertical Support Failure*

Single wall failure describes a mode where a single supporting wall is destroyed, but the slab or the roof above remains unaffected, thus becoming a threat more to people outside than inside the building (witnessed in L'Aquila 2009 and Christchurch 2011) (Fig. 2.10).

Fig. 2.9 A house has it roof collapsed after an earthquake in Nevada, 2008 (Credit: EERI, Photographer: Steven Bartlett and Chris Pantelides)

Fig. 2.10 Single wall failure with an intact roof as witnessed in L'Aquila region, Italy after the 2009 event (*Source* Author)

Fig. 2.11 Multiple wall failures resulting in an overhang collapse observed after the Athens 1999 event (*Source* Pomonis 2000) and an unreinforced masonry (URM) collapse after the Christchurch earthquake of 2011 (*Source* Flickr)

2.3.3 Multiple Vertical Supports Failure

With the loss of more than one vertical support, whether this is a wall or a column, an overhang or cantilever may form and the roof may partially or completely collapse as shown in Fig. 2.11. The floors of the building shown in Fig. 2.11 are clearly tilted which would significantly reduce the chances of survival in such a building, although there are clear visible voids left by the support of the remaining vertical members.

When the height of the collapsed buildings increases, the failure of more than one vertical structural element (a load-bearing wall in masonry or columns in reinforced concrete frames) often results in failure and collapse of floors as well; this will contribute to more entrapment and deaths amongst occupants. This is often called gravitational or progressive collapse mechanism, as shown in Fig. 2.12.

2.3.4 Soft-Storey Collapses

Soft-storey collapses, commonly encountered in reinforced concrete frame buildings where the beam-column or slab-column (in case of flat slab structures) connection completely fails, are characterised by the failure of particular floors which collapse almost uniformly (within the plan of the structure). The building can be mostly preserved in its form and structure but reduced in height, except in worst cases of complete pancake collapse. In Schweier's catalogue, seven types of pancake collapses are identified, depending on the part of the building that is damaged and if one or more stories are affected. Figure 2.13 shows an image of soft-storey reinforced concrete frame collapse witnessed in the 2009 L'Aquila earthquake.

It is not only reinforced concrete frames (in the present day, one of the most common types of structures for housing in the world) that can exhibit soft storey collapses due to earthquake ground shaking, as shown in Fig. 2.14, after the Kobe

Fig. 2.12 A multiple RC column failure of a seven storey RC frame building in L'Aquila after the earthquake in 2009 (*Source* Author)

Fig. 2.13 A soft-storey collapse of a reinforced concrete building after the 2009 L'Aquila earthquake (*Source* Foulser-Piggott 2014)

Fig. 2.14 Photograph shows a single soft storey collapse of a steel frame building (City Hall) after Kobe 1995 (Credit: Courtesy of Earthquake Engineering Field Investigation Team (EEFIT), UK)

earthquake of 1995, a steel frame building was also found to have failed with a single soft-storey collapse.

In the context of survivability inside these collapsed buildings, an example from the 2004 Moroccan earthquake at Al-Hoceima illustrates this well. As shown in Fig. 2.15, even within a complete floor collapse, there are available spaces though in this case, perhaps only if one was by the shelfing unit (circled in red).

Fig. 2.15 Soft storey collapse of a reinforced concrete frame building in after 2004 Moroccan earthquake at Al-Hoceima. Zooming into the building as shown on the right, the chances of survival clearly depend on the available space. (*Source* Flickr)

Fig. 2.16 Complete failure of a non-ductile reinforced concrete frame with brick infill walls structure in Ano Liosia during the Athens earthquake of 1999 (L) and (R) collapse of unreinforced masonry buildings after the Bam 2003 earthquake (*Source* Pomonis and Flickr)

2.3.5 Complete Failure of Structural Elements (Pancake Collapse)

Pancake collapses result from the failure of all structural elements. Depending on the building material and structural system, this damage type can be described as 'heaps of debris' or a dramatic reduction of height ('sandwich' type behaviour) but with intact roofs, where all vertical elements have failed. Figure 2.16 shows the complete failure of a reinforced concrete building after the Athens earthquake of 1999 where all structural components have failed. The right hand figure shows the failure in weak masonry, where debris heaps are formed from disintegrated masonry blocks (2003 Bam earthquake).

2.3.6 Overturn/Toppling

Overturn or toppling collapses consist of the damage types related to a loss of a vertical structural support at the lowest or lower floors of a building, usually multi-storey. This is different from soft storey collapses as the entire storey has not failed and therefore the upper floors rotate and collapse at the location of the failed element as shown in Fig. 2.17. For a complete overturn collapse, the building is still intact but collapses outside of the footprint area on one of the sides or corners, possibly onto adjacent buildings.

Buildings which collapse and overturn could also separate, where the lower part of the building is still at its original position, but the upper part lies separately next to it. With this collapse mechanism, the collapsed building retains spatial integrity but fatalities and injuries are caused by the impact and where the building separates. The most recent building to exhibit this type of collapse mechanism was the Alto Rio building (Fig. 2.18) in the 2010 Concepción, Chile earthquake; the higher floors fell over a long distance and separated upon impact, though most fortunately,

Fig. 2.17 Overturning collapses observed after the Kobe earthquake of 1995 (L) (Credit: EERI) and overturning collapse witnessed in Taiwan after the Chi Chi earthquake (R) (*Source* F. Naiem)

Fig. 2.18 The collapse of the Alto Rio building in the 2010 Concepción, Chile earthquake (Credit: Wikipedia)

only eight people died, further demonstrating the importance of spatial integrity in fatality generation.

Table 2.1 shows illustrations of building damage categories and subtypes with an assessment of the volumetric reduction of the collapse mechanisms.

Based on an evaluation of possible collapse mechanisms and a careful examination of the least amount of volume reduction that causes fatalities, the definition used in this book for the assignments of fatality rates from a collapsed building is as follows:

Table 2.1 Collapse mechanisms of different building types encompassed in D5 (collapse) with increasing volumetric reduction

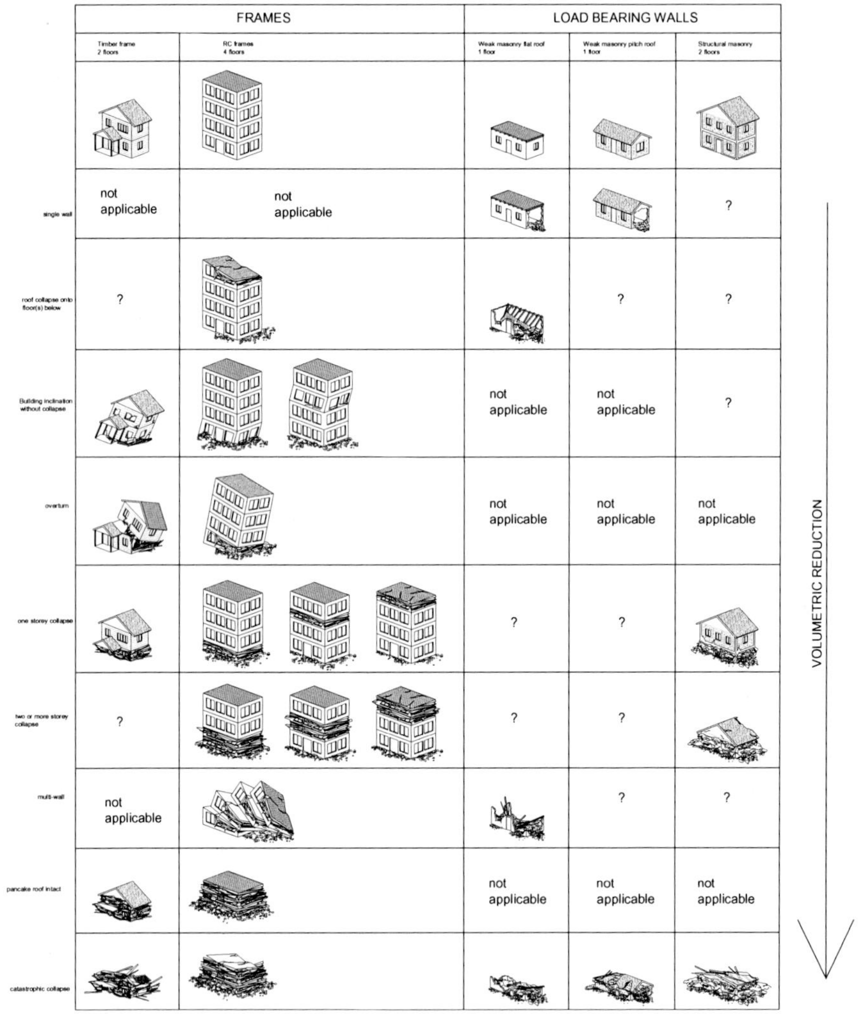

	FRAMES		LOAD BEARING WALLS		
	Timber frame 2 floors	RC frames 4 floors	Weak masonry flat roof 1 floor	Weak masonry pitch roof 1 floor	Structural masonry 2 floors
single wall	not applicable	not applicable			?
roof collapse onto floor(s) below	?			?	?
Building inclination without collapse			not applicable	not applicable	?
overturn			not applicable	not applicable	not applicable
one storey collapse			?	?	
two or more storey collapse	?		?	?	
multi-wall	not applicable			?	?
pancake roof intact			not applicable	not applicable	not applicable
catastrophic collapse					

VOLUMETRIC REDUCTION

At least 10 % volume reduction at any floor level from whatever cause or mechanism of failure.

The collapse mechanisms would depend on building typologies and the characteristics of the ground motion. Future developments into assessing damage states may call for separate definitions of the term "collapse", according to its lethality potential associated with the mechanism of failure. The author proposes a fatality-centric damage categorisation as the way forward.

2.4 Proposing a Range for Fatality Rates in a Collapsed Building

Due to the variation in collapse mechanisms even for the same building types, it was decided that unless evidence suggests otherwise, in this study, a range would be proposed for the percentage of the number of occupants within a collapsed building who may be killed. Apart from variability in global adaptations of building types changing its collapse mechanisms, the degree of collapse and therefore resulting volumetric reduction also depends on ground shaking severity (Coburn et al. 1992). A small proportion of weak masonry buildings may collapse at intensity VI but many more would at IX and at this intensity level, as these collapses may be more catastrophic. With a significant reduction in internal spaces, collapses at higher intensities may therefore be more deadly. This is hard to quantify with precision but is inherent within the observational data the assigned fatality rates are based on.

The use of a broad description for collapse for estimating fatality ratio is a crucial assumption of this method. In order to attempt to overcome this limitation, where the data will allow, fatality rate assignments are associated with identifiable building types and their collapse mechanisms, rather than for the generic building type.

Further work is being carried out by the author to augment the fatality rates proposed in this book with probability density functions derived from actual distributions of fatality data from recent earthquakes. It is hoped that this methodology can be further developed when more data on fatalities become available for past and future events.

References

Coburn AW, Spence RJS (2002) Earthquake protection, 2nd edn. Wiley, Chichester, p 436

Coburn AW, Spence RJS, Pomonis A. (1992) Factors determining casualty levels in earthquakes: mortality prediction in building collapse, In: Proceedings of the 10th world conference of earthquake engineering, Madrid, Spain, 19–25 July, 1992

EEFIT (Earthquake Engineering Field Investigation Team) (2009) The L'Aquila, Italy Earthquake of 6 April 2009: a field report by EEFIT: earthquake engineering field investigation team, Institution of Structural Engineers, London

Foulser-Piggott R. (2014) 2009 L'Aquila Italy: geoarchive photgraphic study (CAR) – Location 96, EEPI Map, Cambridge Architectural Research Ltd. http://snapandmap.com/photo/13312

Galetzka J, Melgar D, Genrich JF et al. (2015) Slip pulse and resonance of the Kathmandu basin during the 2015 Gorkha earthquake, Nepal. Science 349(6252):1091–1095. doi:10.1126/science.aac6383

Jaiswal K, Wald DJ, Hearne M (2009) Estimating casualties for large earthquakes worldwide using an empirical approach: US geological survey open-file report, 2009–1136, 83 p. http://pubs.usgs.gov/of/2009/1136/

Marano KD, Wald DJ, Allen TI (2009) Global earthquake casualties due to secondary effects: a quantitative analysis for improving rapid loss analyses. Nat. Hazards 49. doi:10.1007/s11069-009-9372-5

NIBS-FEMA (National Institute of Building Sciences-Federal Emergency Management Agency) (2006) HAZUS-MH MR2 Technical Manual

Okada S, Takai N (2000) Classifications of structural types and damage patterns of buildings for earthquake field investigation. In: Proceedings of the 12th world conference on earthquake engineering, Auckland, New Zealand, 30 January–4 February, 2000–6, 2004. Paper No. 705

Pomonis A (2011) Director, Cambridge Architectural Research Ltd

Schweier C, Markus M (2006) Classification of collapsed buildings for fast damage and loss assessment. Bull Earthq Eng 4(2):177–192

So EKM (2009) The assessment of casualties for earthquake loss estimation: Cambridge, University of Cambridge, PhD dissertation

Yong P, Yu J (2008) Wenchaun Reconnaissance Report #2, NZSEE, Wellington, New Zealand. http://www.nzsee.org.nz/projects/past-earthquakes/2008-wenchuan-earthquake/wenchuan-reconnaissance-report-2/

Chapter 3
Supporting Literature for Deriving Fatality Rates

One of the main issues of casualty modeling has been a lack of good empirical data from past events from which to derive realistic fatality rates. The quality and consistency of data are also major concerns; hence in collating data for this book, a concerted effort was made in obtaining and assembling global casualty information from recent earthquakes. The levels of resolutions of data found were as follows:

1st level: global data on the overall fatality count per event, some with secondary information on the causes of the deaths (building collapse, slope failure, tsunami, fire following). The information for this comes directly from the USGS PAGER-CAT (Allen et al. 2009).

2nd level: fatality numbers over population at particular geographical units that contain several population centres (i.e., county, district).

3rd level: damaged building types per village/city/neighbourhood and number of people killed overall in each building type (often paired with an approximation of population per building) that can be linked to the level of ground shaking intensity and exclude life losses due to secondary hazards. Such data are rare. Although for some events we know that almost all the deaths were related to a particular type of structure, e.g., adobe in Bam 2003; rubble stone in Maharashtra 1993; Reinforced Concrete (RC) frame (pre-1985) in Athens 1999. Events where one can assign deaths into a specific type of structure with any certainty are limited and are mostly where a few catastrophic building collapses dominate, e.g., Northridge 1994 and Christchurch 2011.

4th level: actual building-by-building survey of structure types and damage levels corresponding with the number killed (and injured) amongst the known number of occupants at the time of an event. There were only three surveys used in this book (So 2010).

Since one of the main assumptions in this methodology is that fatalities must correspond to collapses of buildings, the amount of data are reduced by filtering to only levels 3 and 4. However, some inferences were made to information from levels 1 and 2, as these set the scene for earthquake fatalities and give the numbers vital relevance and background, especially in terms of other contributory factors, like time of day.

In thoroughly reviewing earthquake casualty information from the past 40 years, the following 25 events shown in Table 3.1 were evaluated in detail in the

E. So, *Estimating Fatality Rates for Earthquake Loss Models*,
SpringerBriefs in Earth Sciences, DOI 10.1007/978-3-319-26838-5_3

Table 3.1 List of earthquakes studied for fatality rate assignment

	Event name	Country	Year	Month	Day	Time (local)	Mag	No. of fatalities	Dominant building type
1	Dasht-e-Bayaz	Iran	1968	8	31	14:47	7.2	12,100	Adobe mud brick
2	Karnaveh	Iran	1970	7	30	4:52	6.8	200	Adobe mud brick
3	Ghir	Iran	1972	4	10	5:37	7.1	5,374	Adobe mud brick
4	Guatemala	Guatemala	1976	2	4	3:05	7.5	22,778	Old adobe
5	Friuli	Italy	1976	5	6	21:06	6.9	989	Stone masonry
6	Tabas	Iran	1978	9	16	19:36	7.4	20,500	Adobe mud brick
7	Irpinia	Italy	1980	11	23	19:34	6.9	3,000	Stone masonry with wooden diaphragms
8	Spitak	Armenia	1988	12	7	11:41	6.9	25,000	Precast concrete frame buildings
9	Latur	India	1993	9	30	3:55	6.2	9,748	Mud mortar stone masonry
10	Northridge	USA	1994	1	17	4:31	6.7	16	Multifamily building (three-storey timber and RC)
11	Kobe	Japan	1995	1	17	5:46	6.8	6,434	Old wooden housing with heavy tiled roofs
12	Aegion	Greece	1995	6	15	3:16	6.4	26	Two mid-rise RC MRF collapses
13	Neftegorsk	Russia	1995	5	29	0:10	7.6	1,995	17 mid-rise RC collapses
14	Kocaeli	Turkey	1999	8	17	3:02	7.4	17,479 (official)	RC 2–4 soft storey
15	Athens	Greece	1999	9	7	14:56	6.0	132	27 collapsed low to mid-rise RC MRF with URM infill
16	Chi Chi	Taiwan	1999	9	20	1:47	7.6	2,100	Mud brick/high-rise RC
17	Bhuj	India	2001	1	26	8:46	7.6	13,830	Adobe and RC MRF
18	Bam	Iran	2003	12	26	5:26	6.6	31,000	Mud mortar adobe
19	Niigata	Japan	2004	10	23	17:56	6.8	35	Timber
20	Kashmir	Pakistan, India	2005	10	8	8:50	7.6	86,000	Stone katcha

(continued)

Table 3.1 (continued)

	Event name	Country	Year	Month	Day	Time (local)	Mag	No. of fatalities	Dominant building type
21	Yogyakarta	Indonesia	2006	5	27	5:54	6.3	5,749	Stone, timber truss roof
22	Pisco	Peru	2007	8	15	18:40	8.0	519	Adobe, quincha roofs
23	L'Aquila	Italy	2009	4	6	3:32	6.3	308	Stone masonry, one catastrophic RC collapse of student dormitory
24	Chile	Chile	2010	2	27	3:34	8.8	521	Tsunami
25	Canterbury	New Zealand	2011	2	22	12:54	6.3	181	Two catastrophic collapses of 4-6 RC frame, brick façade debris onto roads

assessment of fatalities. It is important to note that this review includes earthquakes that have caused 27 % of global life loss due to building collapses from ground shaking since 1970.

References

Allen TI, Marano KD, Earle PS, Wald DJ (2009) PAGER-CAT: A composite earthquake catalog for calibrating global fatality models. Seismol Res Lett 80(1): 57–62, doi:10.1785/gssrl.80.1.57

So EKM (2010) Challenges in Collating Earthquake Casualty Field Data: Chapter 16 *in* Spence R, So E, Scawthorn C, eds, Human Casualties in Earthquakes: Progress in Modelling and Mitigation. Springer, The Netherlands. ISBN-10: 9789048194544

Chapter 4
Assignments of Judgment-Based Fatality Rates

The following chapter discusses the assignments of judgment-based fatality rates. The reference earthquakes and data used to derive a set of proposed fatality rates given the collapses of different structural types are systematically examined.

A simple classification of building stock is used which follows a conventional approach of classification into vulnerability classes based on the principal materials and load-bearing structural system (Spence et al. 2008). The classification needed to be sufficiently detailed to account for real variations in performance, but without large numbers of separate classes. This system allows for two separate timber classes, three masonry classes, 14 reinforced concrete classes, seven steel or metal frame classes making a total of 26 separate classes (mobile homes have been omitted).

Reinforced concrete structures are subdivided by frame and shear-wall structures, by three age categories (related to earthquake code developments) and three height classes, while steel structures were divided between moment-resisting and braced frames and three height classes. The full system is described in Table 4.1. The aim of this study is to find fatality data for each of these classes. However, as expected there are gaps as where there have been difficulties in locating studies or simply for modern code structures, they have either not been subjected to earthquakes or caused fatalities.

For the building types examined, studies of the collapse mechanisms due to ground shaking were carried out. A study of the failure mechanisms is of significant value as victims are generally killed by:

- crushing or suffocation under collapsed structural elements, or
- asphyxiation by the volume of dust generated by the collapse or
- delay in being rescued.

The amount of space (volume) available for occupants when trapped but not killed and of course the speed and ability of search and rescue teams would have an impact on the survivability of a victim in a collapsed building. It is worth noting that an increased death toll due to collapse is likely when the proportion of collapsed structures exceeds a certain threshold e.g., 30 % of total building stock, to account for limitations of search and rescue capabilities (FDMA, oral commun. 2008). These latter factors are much harder to quantify and therefore we concentrate

E. So, *Estimating Fatality Rates for Earthquake Loss Models*,
SpringerBriefs in Earth Sciences, DOI 10.1007/978-3-319-26838-5_4

Table 4.1 The building classification system used in this study

	Material	Structural system	Building class details	Design code compliance	Height sub-class
1	Timber	Light timber construction	Stud walls; Small post and beam; Plywood sheathing; smaller size (<500 m^2)	–	Low-rise (1–3 floors)
2	Timber	Heavy timber construction	Thicker post and beam; masonry infill; usually large floor area (>500 m^2)	–	Low-rise (1–3 floors)
3	Masonry	Weak masonry load bearing walls	Adobe, rubble, irregular stone	–	Low-rise and medium-rise (1–7 floors)
4	Masonry	Unreinforced masonry load bearing walls	Brick, concrete block, hewn regular stone, large stone with timber or concrete or metal deck floor diaphragms	–	Low-rise and medium-rise (1–7 floors)
5	Masonry	Structural reinforced masonry walls	Reinforced brick masonry, confined masonry, dual masonry wall with metal or RC frame system with timber or concrete (incl. Precast) or metal deck floor diaphragms	–	Low-rise and medium-rise (1–7 floors)
6	Reinforced concrete (RC)	Frame	Cast-in situ or precast RC frame with infill walls of usually unreinforced masonry incl. Flat slab system	Pre-code or noncompliant to code	Low-rise (1–3 floors)
7	RC	Frame	Ditto	ditto	Medium (4–7 floors)
8	RC	Frame	Ditto	Ditto	High-rise (>7 floors)

(continued)

Table 4.1 (continued)

	Material	Structural system	Building class details	Design code compliance	Height sub-class
9	RC	Frame	Ditto	Early code (equival. static force design)	Low-rise (1–3 floors)
10	RC	Frame	Ditto	Ditto	Medium (4–7 floors)
11	RC	Frame	Ditto	Ditto	High-rise (>7 floors)
12	RC	Frame	Ditto	Modern code (ductility, spectral demand design)	Low-rise (1–3 floors)
13	RC	Frame	Ditto	Ditto	Medium (4–7 floors)
14	RC	Frame	Ditto	Ditto	High-rise (>7 floors)
15	RC	Shear wall	Cast-in situ or precast RC shear wall systems	Regardless of code vintage	Low-rise (1–3 floors)
16	RC	Shear wall	Ditto	Ditto	Medium (4–7 floors)
17	RC	Shear wall	Ditto	Ditto	High-rise (>7 floors)
18	RC	Tilt-up and (or) long span incl. precast	Precast concrete tilt-up walls with timber or steel truss roofing, often long span	Regardless of code vintage	Low-rise (1–3 floors)
19	RC	Dual	Combination of frame and shear wall construction potentially with mixed precast and cast-in situ elements	Regardless of code vintage	Medium-high rise (>4 floors)
20	Steel	Moment resistant frame	Steel frame with various types of lightweight sheathing or with heavier infill walls incl. RC shear walls	Regardless of code vintage	Low-rise (1–3 floors)

(continued)

Table 4.1 (continued)

	Material	Structural system	Building class details	Design code compliance	Height sub-class
21	Steel	Moment resistant frame	Ditto	Ditto	Medium (4–7 floors)
22	Steel	Moment resistant frame	Ditto	Ditto	High-rise (>7 floors)
23	Steel	Steel braced frame	Concentrically or eccentrically braced steel frame with various types of lightweight sheathing or with heavier infill walls incl. RC shear walls	Regardless of code vintage	Low-rise (1–3 floors)
24	Steel	Steel braced frame	Ditto	Ditto	Medium (4–7 floors)
25	Steel	Steel braced frame	Ditto	Ditto	High-rise (>7 floors)
26	Steel	Light metal frame	Small factory, warehouse, shed type	Regardless of code vintage	Low-rise (1–3 floors)

on assessing the failure mechanisms to help understand the lethality potential of collapsing buildings.

In the subsequent sections, each of the building types shown in Table 4.1 is taken in turn. The building class will first be described, followed by an assessment of possible failure mechanisms based on actual collapses from recent events. Disaggregating the data by following failure-collapse mechanisms allow for more flexibility and control and justification for fatality rate assignments. As discussed in the previous chapters, since the definitions of collapse vary, this approach also allows the user to assign a fatality rate to a corresponding volume loss which can range from 10 % (partial collapse but still D5) to 90 % (worst types of pancake collapse RC or heavy roofed weak masonry). The study of collapse patterns is also useful to inform relativities in fatality rates for classes for which limited casualty data are available but photos are more plentiful. Based on an assessment of the 25 events of most significance in the last 40 years as set out in Table 3.1, evidence of these collapses from past events and how they relate to the lethality potential in buildings are reviewed. Lastly, ranges of fatality rates are assigned to the investigated building types stating the corresponding volumetric reduction of the collapsed structure, based on direct observations or inferred from collected evidence.

4.1 Timber

4.1.1 Light Timber (TL)

Light timber housing refers to wood frame construction typically used for single-family houses very widely in the USA, New Zealand and Japan. These houses offer versatility and are often adapted according to climatic conditions, most notably in Japan with heavy roofs for typhoons. The two types of timber have been further divided according to the weight of the roof (light and heavy roofing). As witnessed in recent Japanese events (Kobe 1995 and Niigata 2004), the older heavy roofed structures have collapsed with a greater volume loss and contribute to increased loss of life compared to lighter roof timber structures (Okada 1996).

In these three countries the majority of the population live in timber framed residential housing. In the Christchurch earthquake of 2011, only two of the 181 deaths were attributed to collapses of timber homes, although 1,630 timber homes were red-tagged (mostly due to failure related to soil liquefaction). This would suggest extremely low volume losses but it is also related to easier extrication associated with these generally light structures, leading to low lethality within these structural types.

The photographs in Fig. 4.1 show collapses of this type of housing (TL). The photograph on the left shows the typical collapse mechanism observed after the Kobe earthquake in Japan in 1995 where the first floor of the residential house is crushed completely by the heavy roof, the load of the second floor, as well as potentially the increased decay of the timber in older buildings. The picture on the right is taken from Northridge 1994 and shows damage to a timber residential structure. The available space for egress, the weight of possible crushing objects and the lack of dust highly increases the survivability in this type of dwelling. The opposite is the case for the old timber dwellings in Kobe, where volume loss, dust from the heavy tiled roof (for increased typhoon resistance) and the layer of mud below the tiles (for thermal insulation) contributed towards a much greater lethality potential for the same building material category.

Fatality rates (FR) of 0.25–3 % have been assigned for these structures, inferred from earthquake fatality data collected from the USA, New Zealand and Japan.

The values in Table 4.2 should be compared to the greater values in HAZUS (NIBS-FEMA 2006), which uses 5 % for small and 10 % for big wooden structures. This difference can be attributed to recent evidence from the New Zealand events and also since the Kobe earthquake of 1995, there has been an on-going nationwide retrofit and reconstruction program by the Central Disaster Management Council (Suganuma 2006). By 2003, 3,500 units of old timber housing, the main contributor to the circa 4,950 deaths in the earthquake related to building collapse[1] have been reinforced, though the progress has been slow.

[1]The remaining deaths include approximately 500 deaths due to fires, 34 landslide deaths and 932 post-event deaths.

Fig. 4.1 (*L*) Typical collapse mechanism of timber houses in the Kobe earthquakes where the entire ground floor is crushed (*R*) and typical damage observed after the Northridge earthquake in 1994). (Credit: NISEE, University of California, Berkeley, and FEMA, Photographer: Andrea Booher)

Table 4.2 Fatality rate assignments for collapsed light timber frame structures

Timber	Typical volume loss (%)	Fatality rates (% of occupants)	Earthquake source
Light timber with light roofs	<10	0.25–1	US and NZ earthquakes in 1980–2010s
Light timber with heavy roofs	>50	0.75–3	Collapses of old Japanese timber housing (from FDMA)

4.1.2 Heavy Timber (TH)

Heavy timber frame constructions are more common in Northern Europe, especially in Portugal and parts of Italy. The Portuguese Pombalino buildings have existed since the great 1755 Lisbon earthquake and are early forms of braced frame timber construction, where the wooden frames are infilled with brick masonry or adobe blocks (Cardoso et al. 2011). In Asia, similar forms of earthquake-resilient vernacular construction can be found as dhajji-dewari in India and himis in Turkey (Fig. 4.2).

There has been little evidence of collapses of these types of vernacular housing which are often promoted as earthquake resistant structures (Langenbach 2009). Therefore, fatality rates for collapsed heavy timber frame structures have been proposed to be between light weight timber buildings and unreinforced masonry structures since the failure of masonry infill walls have a potential to kill a larger proportion of the occupants (Table 4.3).

Fig. 4.2 Photographs of a Pombalino building (Credit: EERI) and a dhajji-dewari building in India (*Source* Durgesh Rai, Indian Institute of Technology, Kanpur)

Table 4.3 Fatality rate assignments for collapsed heavy timber frame structures

Timber	Typical volume loss (%)	Fatality rates (% of occupants)	
Heavy timber with light roofs	<10	0.5–1	Inferred from collapses of light timber and URMs
Heavy timber with heavy roofs	>50	2–3	

4.2 Masonry

For the generic building class of masonry, there are three main types namely *weak*, *load bearing* and *reinforced masonry*. Further divisions have been added to account for the variability of masonry structures around the world, adapting for local cultural and climatic factors. Through recent field surveys and studies, it was found that these changes to the buildings' attributes play an important role in the failure mechanism and therefore the survivability of occupants (So 2009). A thorough assessment of this class of structures has been made possible as observations from several recent earthquakes from Bhuj, India in 2001 to the 2009 event in L'Aquila, Italy have helped improve our understanding of fatalities in masonry buildings.

4.2.1 Weak Masonry

Weak masonry consists of adobe and irregular rubble stone structures, usually set in mud or lime mortar. These houses are typically single to two-storey high and house on average of 4–7 people (So 2009) in countries of the developing world, while in

Europe their occupancy is often quite low (e.g., less than 1.1 person per building in the case of Greece). The types of weak masonry have been further divided by roof types. Recent surveys in Peru and Pakistan have shown that the influences of roofing material and its weight are significant (So 2009).

The actual constitution of the masonry and the type of roof played a part as apparent in 2003 Bam (Iran), where people not only died as a result of the weight of the falling walls and roofs but many more did not survive due to asphyxiation (Kuwata et al. 2005). This could partly help explain the difference between the 10 % fatality rate evident in completely destroyed adobe housing in the 2007 Pisco (Peru) earthquake where the roofs consisted of lightweight matted bamboo and the 40–60 % fatality rates witnessed in Iranian earthquakes of the 1970s (e.g., 1972 Ghir, 1978 Tabas earthquakes) and also the high lethality in the 1970 Ancash (Peru) event, compared with the Pisco event where the egress by residents may have reduced lethality, overall.

In Peru, although in both regions adobe houses were dominant, in the 1970 Ancash earthquake much of the affected region was in the Andes Mountains and the roofs were covered with heavy tiles; while in Pisco which is by the sea and in a desert area, the typical roofing was much lighter. In the 1970 event, the coastal city of Chimbote experienced shaking of intensity VIII and most of its adobe buildings were destroyed (Berg and Husid 1971) with the city losing 0.6 % of its 117,500 people (Plafker et al. 1971). In Huaraz (at an elevation of 3,050 m) the shaking intensity was VII–VIII but almost all the adobe houses in the southern half of the city were destroyed and the city lost 26 % of its 65,300 people (Plafker et al. 1971). It is worth noting that amongst this devastation, none of the well-constructed RC buildings suffered more than moderate damage due to shaking (Berg and Husid 1971). The photos below show the differences in roofing between weak masonry houses in Pisco and Ancash in Peru (Fig. 4.3).

Earthquakes in China, Indonesia, India and Pakistan have been the main references for the assessment of fatality rates for irregular rubble stone houses shown in Table 4.4. Once again, the observed failures were mostly residential houses of one to two storeys made with locally found stones in the rural setting. These may be

Fig. 4.3 Single storey adobe residence with woven bamboo roof cover found in the coastal regions of southern Peru (*L*) (*Source* Author) and adobe house with heavier roof found in the Andean regions of Peru (*R*) (*Source* Flickr)

Table 4.4 Fatality rate assignments for collapsed weak masonry

Weak masonry	Typical volume loss (%)	Fatality rates (% of occupants)	Reference earthquakes
Adobe light roof	<50	5–15	Guatamala 1970, Pisco 2007
Adobe heavy roof	>75	20–90[a]	Iran 1970–2003, Ancash, 1970
Irregular stone with wooden pitched roofs (low-rise)	40–60	5–20	Irpinia 1980, Wenchuan 2008, Yogyakarta 2006
Irregular stone low-rise concrete slab roofs	>70	10–40	Kashmir 2005

[a]A wider range applies to URM as so many factors affect the FR and volume loss would range from 10 to 80+ % depending on intensity and other factors

irregular in shape and size and can be very poorly joined with weak mortar. The percentage of occupants killed due to collapse of these structures was found to depend on the material and shape of the roofs. In China (Wenchuan 2008), the pitched tile roofs were supported by a wooden truss. These types of structures were found to be less lethal and the fatality rates ranged from 5–15 % in the casualty data as attained from the Yogyakarta earthquake (So 2009), although still failing at low intensities. In the Wenchuan earthquake, some buildings failed completely at intensities as low as VII due to the poorly connected stone walls (Sun and Zhang 2010).

By contrast, the Kashmir (Pakistan) event of 2005 revealed mixed material construction where stone masonry was used to support flat concrete slab roofs. As the walls failed, the heavy roofing structure proved much more lethal. The proportion of occupants killed in these types of housing from the earthquake was found to range from 18–27 % (So 2009) and could be much higher since only surviving households were interviewed in So's survey. A summary of the suggested rates for weak masonry are shown in Table 4.4.

4.2.2 *Unreinforced (Load Bearing) Masonry*

Unreinforced masonry (load bearing) can be divided into two categories, representing the different lethality potential of buildings, i.e., with wooden and concrete flooring. The data supporting the derivation of these fatality rates are from Europe, where wooden floors are common (e.g., Italy) and Asia (Indonesia, Taiwan and China), where concrete floors are more typical.

Unlike weak masonry structures, load bearing masonry structures can be up to seven storeys and the height of a building does affect its lethality potential (De Bruycker et al. 1985) with an increase in fatality rates with increasing height of the building. Figure 4.4 shows two typical examples of this building type.

Fig. 4.4 Collapse of low-rise load bearing brick masonry buildings after the 2006 Yogyakarta, Indonesia (*L*) and mid-rise collapse in 2009 L'Aquila, Italy earthquakes (*Source* Author)

Fig. 4.5 Total collapse of a dormitory, 2008 Wenchuan (Credit: EERI, Photographer: Dennis Lau)

Although two walls have completely collapsed, the support system of the other walls and the light weight wooden truss and tiles provided life safety for the inhabitants in the left hand picture. By contrast, the load bearing walls of a three-storey building in Onna, Italy, again losing two walls proved more lethal as the failed floors fell on the inhabitants on the ground floor (EEFIT 2009).

Examining collapses of low to mid-rise masonry buildings with wooden floors in Italian earthquakes revealed typical volume losses of completely damaged buildings

Table 4.5 Fatality rate assignments for unreinforced (load bearing) masonry

Unreinforced masonry	Typical volume loss (%)	Fatality rates (% of occupants)	Reference earthquakes
European (wooden floors)	>30	3–12	Italy 1970s–1990s
Asian (concrete floors)	>50	10–25	Chi Chi 1999, Wenchuan 2008

of >30 %, with a fatality rate of 9–12 %. An inference has been made that the majority of load bearing masonry buildings in Italy have wooden floors. This is based on a analyses of 115,000 masonry buildings from the Italian post-earthquake damage database showing that 64 % had wooden and 36 % had RC floors (Rota et al. 2008).

As for masonry buildings with concrete flooring, evidence from the 2008 Wenchuan earthquake where precast hollow-core planks were used for the floor systems, were examined. Observations of the damage suggest that there were no connections between the precast panels and the supporting brick walls. Typically out-of-plane failures of top storeys due to inadequate lateral restraint were found as shown in Fig. 4.5.

The fatality rate amongst these collapsed structures ranged from 10–25 % with volume reductions greater than 50 %, as inferred from damage studies carried out after the Wenchuan event (Table 4.5).

4.2.3 *Reinforced Masonry*

In reinforced masonry, bars or steel mesh are embedded (in mortar or grout) in holes or between layers of masonry bricks, creating a composite material acting as a highly resistant and ductile wall or wall system. Such reinforcement can be present in both the vertical and horizontal directions. In certain regions special stone systems are developed where shaped (e.g., interlocking) building stones are formed out of concrete; these also perform very well. Another efficient system is known as grouted masonry, comprising walls consisting of an outer and inner brick shell, connected with a concrete core vertically and horizontally reinforced as described in the European Macroseismic Scale 1998 (EMS-98) though experience with this form of construction is limited at present. Therefore fatality rates for collapsed reinforced masonry structures have been proposed to be between unreinforced masonry and confined masonry buildings, and again concrete floored structures have been assigned higher rates reflecting a higher potential to kill with the collapses of concrete slabs (Table 4.6).

Table 4.6 Fatality rate assignments for reinforced masonry

Reinforced masonry	Typical volume loss (%)	Fatality rates (% of occupants)	Reference earthquakes
Reinforced masonry (low-rise and wooden floors)	>10	2–8	Inferred
Reinforced masonry (mid-rise with concrete floors)	>20	15–40	

4.2.4 *Confined Masonry*

Over the last 30 years, confined masonry construction has been practiced in many regions, among others in Mediterranean Europe (Italy, Slovenia, and Serbia), Latin America (Mexico, Chile, Peru, Argentina, and other countries), the Middle East (Iran), south Asia (Indonesia), and the Far East (China). In confined masonry construction, the masonry walls carry the seismic loads and the concrete columns are used to confine the masonry walls.

By and large, confined masonry buildings performed very well in providing life safety in the 2010 Maule (Chile) earthquake. Most one and two-storey single-family dwellings did not experience any damage, except for a few buildings which suffered moderate damage. There were as noted by Brzev et al. (2010), two 3-storey confined masonry buildings that collapsed in Constitución and Santa Cruz. Most damage in these two buildings was concentrated to the ground floor, with a complete soft-storey failure noted in Santa Cruz as shown in Fig. 4.6. In each block there were 12 units, so at least 20 people were inside the collapsed building at the time of the earthquake. With two deaths, the fatality rate in this building was less

Fig. 4.6 Collapse of a three-storey confined masonry building in Santa Cruz (2010 Maule, Chile) earthquake (*Source* Svetlana Brzev)

than 10 %. The building was assessed by EERI engineers concluding that the failure was attributed to poor quality of construction for both brick and concrete block masonry and the low wall density (less than 1 % per floor); though out of the 28 identical three-storey confined masonry buildings in the complex, only one collapsed (Yadlin, oral commun. 2011).

By contrast, confined masonry buildings which were widespread in Port-au-Prince, Haiti did not conform to code specifications and did not perform well in the 12 January 2010 earthquake, also due to more severe ground motion levels. Tragically, as noted by the EEFIT team, the Haitian version had the outward appearance of confined masonry but had been built without the seismic detailing necessary to provide confinement of the masonry walls, resulting in thousands of catastrophic failures (EEFIT 2010). The confined masonry construction witnessed in Haiti, used in some cases for multi-storey buildings performed no better than its weaker unreinforced masonry counterpart.

In terms of assessing the lethality potential of collapsed confined masonry buildings, as demonstrated in comparing the 2010 Haiti and Chile earthquakes, realising the quality of construction is vital (Fig. 4.7).

Fig. 4.7 Mass destruction of confined masonry building in Port-au-Prince (Credit: Courtesy of Earthquake Engineering Field Investigation Team (EEFIT), UK)

4.2.5 Mixed

Termed "Mixed", this class of buildings has become increasingly common in growing cities in China. Commonly built as a reinforced concrete structure in the first, or two lowest floors with load-bearing masonry walls in the upper floors, these buildings usually have cast-in situ or precast RC slabs and therefore their collapse can be lethal. This mixed structural system was widespread in the region affected by the 2008 Wenchuan earthquake in China and exhibited different vulnerability to RC frame buildings and unreinforced masonry buildings, thus forming a separate structural class in terms of risk assessment and casualty estimation.

The actual performance and therefore the lethality potential of these structures are difficult to assess. However, from events in China, by analysing the failure pattern at different intensities (Sun and Zhang 2010), some assessment of the likely fatality rates in this type of collapsed building has been made. The photograph in Fig. 4.8 shows a hotel with the two bottom floors collapsed in the back of the building. According to engineers investigating these collapses in China (Sun and Zhang 2010), this collapse mode was more common in areas of high intensities in the worst affected counties of Beichuan and Wenchuan, where the stiffer reinforced concrete section of the building 'attracted' more of the imposed seismic forces. In lower intensity areas, it was noted that the collapse rate was very low to null.

Based on these observed trends of failure, the suggested fatality rates for the confined masonry and mixed buildings are tabled below. These have been inferred from collapses with similar volume losses (Turkish reinforced concrete structures). It should be stressed that collapses of confined masonry and mixed structures (not in seismic design codes) have been mainly subject to poor detailing and construction method. Damage of these poorly constructed buildings observed in recent events would suggest at least a pancaked floor with volume reductions of over 30 % (Table 4.7).

Fig. 4.8 Failure mechanisms observed in mixed masonry and RC buildings in Ying Xiu in China (*Source* Sun and Zhang 2010)

Table 4.7 Fatality rate assignments for confined and mixed masonry

Masonry	Typical volume loss (%)	Fatality rates (% of occupants)	Earthquake source
Confined masonry (to code)	<10	2–6	Chile 2011
Low quality	>60	10–40	Haiti 2010
Mixed	>30	15–20	Wenchuan 2008

4.3 Reinforced Concrete

There are 14 reinforced concrete classes specified in the classification system shown in Table 4.1 but in reality, the casualty information that has been obtained for collapsed reinforced concrete buildings is very limited. Most of the modern engineered buildings have not been tested under real earthquake loading and we may have to rely on analytical methods to derive their vulnerability and likely lethality potential in the future. However, should a high-code RC frame building collapse, it is likely that the fatality rate would be less than that of low code reinforced concrete frame buildings as the redundancies built into high code RC structural systems (detailing and increased ductility) should prohibit a catastrophic pancake collapse.

For the current study, in some instances due to a lack of data, case studies of collapses of reinforced concrete buildings not caused by earthquakes have been examined to explore collapse mechanisms and the fatality pattern of these structures. The Sampoong Department Store collapse in Seoul in 1995 and the 1999 collapse of an apartment building in Foggia, Italy have been used as reference. The following sections present the observed building collapses and lethality potential of the different classes of reinforced concrete.

4.3.1 Low to Medium Rise Collapses

For the assessment of fatalities associated with low to medium rise reinforced concrete (RC) collapses, studies of casualties in concrete frame buildings after the Kocaeli earthquake in Turkey in 1999 and Wenchuan earthquake of 2008 (Petal 2004; Sun and Zhang 2010) were consulted. Though it was identified after these events that the failure of most reinforced concrete frame buildings was due to non-conformance with building codes and substandard construction materials, the mechanisms of collapses give plausible modes of beam-column connection failures and therefore potential fatalities. The typical failure mechanism observed in Wenchuan for low-rise reinforced-concrete frames was soft-storey collapses of the lowest floor or floors and a twisted overlying structure as shown in Fig. 4.9.

Fig. 4.9 Failure of a concrete frame building after the Wenchuan Earthquake of 2008 (Credit: EERI, Photographer: Dennis Lau)

Fig. 4.10 Complete failure of a seven-story reinforced concrete frame structure with hollow-brick infill in Golcuk after the Kocaeli Earthquake of 1999 (*Source* NOAA National Geophysical Data Center)

These types of failures resemble the failure of low to medium rise concrete frame buildings after the 1999 Kocaeli earthquake, although with weaker upper columns, whole storey collapses were also seen on the upper floors as shown in Fig. 4.10. The estimated volume loss from this type of collapse mechanism can be well over 50 %, although possibly still providing survival voids as shown in Petal's (2004) study of over 500 households, where the fatality rate in completely collapsed RC buildings was around 13 %.

The survey results from Petal's (2004) study also shows that 50 % of interviewed households in collapsed reinforced concrete buildings were not injured (Table 4.8), although the sample may be skewed due to the need for such a survey to interview members from surviving households.

Table 4.8 Survey results of occupants in totally and partially collapsed low-rise and multi-storey concrete frame buildings, 1999 Kocaeli Turkey earthquake (Petal 2004)

Level of building damage	Total collapse		Partial collapse		Total
Building height (storeys)	5–10	1–4	5–10	1–4	
Death on arrival (DOA)	12.7 (n = 33)	0.0 (n = 0)	2.0 (n = 3)	0.0 (n = 0)	7.0 (n = 36)
Died in hospital	0.4 (n = 1)	0.0 (n = 0)	0.0 (n = 0)	0.0 (n = 0)	0.2 (n = 1)
Hospitalised	3.5 (n = 9)	1.7 (n = 1)	0.0 (n = 0)	0.0 (n = 0)	1.9 (n = 10)
Hospital care: treat and release	5.8 (n = 15)	3.4 (n = 2)	0.0 (n = 0)	0.0 (n = 0)	3.3 (n = 17)
Out-of-hospital care: treat and release	17.0 (n = 44)	8.6 (n = 5)	7.4 (n = 11)	3.9 (n = 2)	12.0 (n = 62)
Injured but no treatment sought	10.0 (n = 26)	5.2 (n = 3)	8.7 (n = 13)	9.8 (n = 5)	9.1 (n = 47)
Not injured	50.6 (n = 131)	81.0 (n = 47)	81.9 (n = 122)	81.0 (n = 47)	66.5 (n = 344)
Total N	N = 259	N = 58	N = 149	N = 51	N = 517

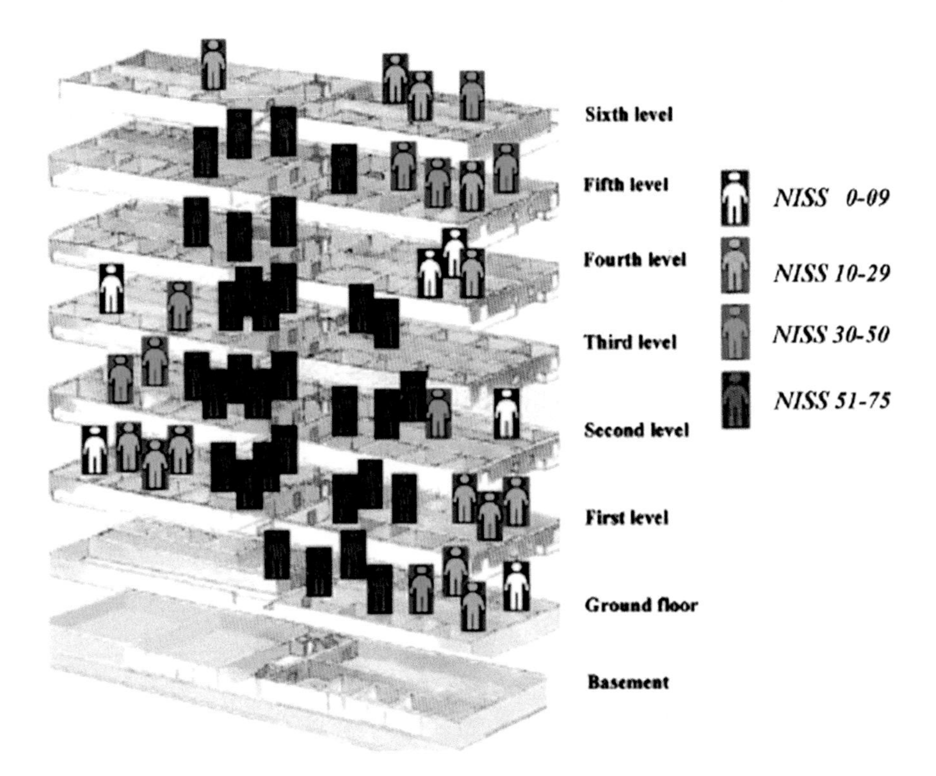

Fig. 4.11 The presumed distribution of the identified 60 victims considering the location of their apartments and their injury severity scores (*Source* Campobasso et al. 2003, reproduced by permission of ASTM International)

Of course, this is only a very small sample and in assessing the reliability of these numbers, we also searched for information on collapses not from earthquakes. In November of 1999, a six-storey apartment building collapsed in Foggia in Italy. The building was a pre-1970s construction and the cause of the collapse was cited as the failure of the building foundations (Campobasso et al. 2003). The six-storey building collapsed, floor upon floor, starting from the centre where the stairways were located. The collapse reduced the building to rubble approximately one storey high, and a gas fire was ignited in part of the ruins (volume loss was close to 90 %). The fatality rate of this completely collapsed building was 88 % (with 61 bodies recovered, 5 still unaccounted for, 1 died in hospital, out of 76 inhabitants).

In addition, the medical post-mortem of this collapse provided a rare example of the distribution of victims within the building with their injury scores and an assessment of the causes of deaths (Fig. 4.11).

The collected evidence and information show that the number of people killed in a collapsed RC low to mid-rise building can range from 5 % to nearly 90 %, depending on the location, failure mechanism and whether there are secondary complications like fire following the main earthquake. This is to be expected as volume loss in collapsed RC buildings can range from less than 10 % to more than 80 %, e.g., in Turkey 1999 many RC frame buildings collapsed but were supported by adjacent buildings that withstood the earthquake thus reaching a much lower volume loss than if they were free standing.

4.3.2 *High-Rise Collapses*

The collapses of pre-1941 gravity-load design high-rise RC frames in the 1977 Vrancea earthquake in Romania was used as the starting point to the exploration of lethality potential of high-rise RC collapses. It was found that out of the 20 collapsed buildings of this class (Georgescu and Pomonis 2011); the overall volume loss was around 75 % with an estimated fatality rate around 60 %. There is high confidence in the volume loss assessment (as photos of all the collapsed high-rise buildings in Bucharest were studied) and moderate confidence in the likely number of people inside the occupied collapsed buildings at the time of the earthquake (9:15 pm local time), as the number of dwelling units for most of the collapsed residential buildings is available, but casualty data are not available on a building by building basis. There is a lower confidence in the fatality rate assigned to these failures (Pomonis, written commun. 2011).

Next, we turned to the overturning collapse (RC shear wall structure) of the Alto Rio in Concepcion, Chile 2010 earthquake, shown in Fig. 4.12. This had a fatality rate of 9 %. The Alto Rio apartment block in Concepción, Chile was a newly constructed 15-storey RC shear wall structure that was partly occupied (many of the

Fig. 4.12 The overturning collapse of the Alto Rio building in the 2010 Concepción, Chile earthquake. The FR was 9 %. (Credit: Wikipedia)

Fig. 4.13 A RC high-rise collapse (overturn) in Bucharest 1977 (a section of Bloc OD16) with a FR ~25 % (Credit: URBAN-INCERC, reproduced with kind permission)

Fig. 4.14 RC shear wall collapse (overturn) example from the 1999 Chi Chi earthquake in Taiwan (Credit: Wikimedia)

apartments were vacant and on the market for sale or rental, awaiting new tenants). This was the only building over three storeys in height that completely collapsed in the area of Concepción during the earthquake. At the moment of the earthquake, there were 87 occupants in the building; there were 8 deaths and 79 survivors, of which 52 persons were able to evacuate the building on their own, and the remaining 27 were rescued from the debris (El Mercurio, 7 March 2010).

By contrast, a similar 11-storey reinforced concrete shear wall collapse (overturn) in Bucharest in the 1977 earthquake (Bloc OD16) exhibited an estimated fatality rate of 20–25 % (shown in Fig. 4.13).

The variation in fatality rate from 9 to 25 to 60 % could be explained by the collapse mechanisms (pancake versus overturning) and the local search and rescue efforts. However more evidence is needed to support these findings and the search continues to seek out unpublished local data for tall building collapses, especially from the Chi Chi, Taiwan 1999 earthquake. Figure 4.14 shows a type of overturning failure that was witnessed after this 1999 event.

Table 4.9 has been adapted and extended from Pomonis et al. (1991) and summarises the lethality of collapses for RC buildings during recent earthquakes which was used as reference to derive the fatality rates for reinforced concrete buildings in Table 5.1.

Table 4.9 Lethality of collapses for RC buildings during recent earthquakes (adapted and extended from Table 3.II, Pomonis et al. (1991)

	Nuevo Leon estate	Medic cent dormitory	General hospital	Juarez hospital	Ruben Darlo centre	Santa Catalina school	Ministry of planning	Apartment building	Sewing factory
Earthquake	Mexico city	Mexico city	Mexico city	Mexico city	San Salvador	San Salvador	San Salvador	Kalamata	Armenia
Time	19/9/1985 (07:17)	19/9/1985 (07:17)	19/9/1985 (07:17)	19/9/1985 (07:17)	10/10/1986 (11:49)	10/10/1986 (11:49)	10/10/1986 (11:49)	13/9/1986 (20:25)	7/12/1988 (20:25)
Vertical structure	RC frame (in situ)	RC frame (in situ)	RC frame (in situ)	RC frame (in situ)	RC frame (in situ)	Hybrid RC, URM	RC frame (in situ)	RC frame (in situ)	Precast concrete frame
Horizontal structure	RC slab and beam	RC slab and girder	RC slab and beam	RC slab and beam	Solid RC flat slab	Wooden floor	Waffle RC flat slab	RC slab and beam	Precast Holl. RC planks
Other struc details	'X' bracing exp. joint	–	–	–	'U' shape	1st fl RC 2nd fl URM	Soft storey	Soft storey	Lack of diaphragm action in slabs
Period of construction	1968–65	1965–1968	Post 1980	1971	Pre 1960			1976	Post 1960
No. of storeys	15	8	6	12	5	2	5	6	–
Previous damage	1970 EQ part repair	–	–	–	1965 EQ unrepaired	–	–	None	None
Cause of collapse	Stiffness discont. in elevation	Soil-struc interaction	Soil-struc interaction	Soil-struc interaction	Torsion and pounding	struc discont. in elevation	Soft storey	Soft storey	Inadeq. joint (col-beam)
Collapse typology	Column fail. and overturning	Column fail. (pancake)	Column fail. (complete coll.)	Column fail. (pancake)	Column fail. (pancake)	2nd fl coll ex. wall coll.	Column fail. (pancake)	Column fail. (pancake)	Col-beam joint failure
Reported occupants	1200	76	471	950	at least 500	at least 50	64	25	212
Killed occupants	468	36	295 (47 missing)	561	at least 300	30	13	6	205
Fatality rate (%)	**39**	**47**	**63**	**59**	**60**	**60**	**20**	**24**	**97**

(continued)

Table 4.9 (continued)

	Nuevo Leon estate	Medic cent dormitory	General hospital	Juarez hospital	Ruben Darlo centre	Santa Catalina school	Ministry of planning	Apartment building	Sewing factory
References	Mexico news	Durkin and Murakami ('88)	Zeballos ('88)	Krimgold ('88)	Olson and Olson ('87)	Anderson ('87) and Durkin ('87)	Anderson ('87)	Anagnostopoulos ('87)	Noji ('90)
	School	Two Apartment buildings	Hotel Eliki	Apartment Building	Apartment Buildings	High-rise Apartments	Student Dormitory	CTV	PGC
Earthquake	Armenia	Armenia	Aeghion	Aeghion	Neftegorsk	Chi Chi	L'Aquila	Christ-church	Christ-church
Time	7/12/1988 (20:25)	7/12/1988 (20:25)	15/6/1995 (3:16)	15/6/1995 (3:16)	29/5/1995 (0:10)	20/9/1999 (1:46)	6/4/2009 (3:32)	22/2/2011 (12:54)	22/2/2011 (12:54)
Vertical Structure	Precast concrete frame	Precast concrete panel	RC Frame (in situ)	RC Frame (in situ)	RC Frame (in situ)	RC Frame (in situ)	RC Frame (in situ)	RC Frame (in situ)	RC Frame (in situ)
Horizontal Structure	Precast Holl. RC planks	Precast Holl. RC planks	RC slab and beam	RC slab and beam	RC slab and beam	RC slab and beam	RC slab and beam	RC slab and beam	RC slab and beam
Other Struc Details	Lack of diaphragm action in slabs	–	–	"L" shape	–	–		–	–
Period of Construction	Post 1960	Post 1960					Post 1960	Post 1970	Post 1970
No. of Storeys	–	2	5	5	5			6	6
Previous Damage	None	None	–	–	–	None	None	Possibly from Darfield eq	–
Cause of Collapse	Inadeq. joint (col-beam)	Inadeq. Panel connections	Struc discont. in elevation	–	Struc discont. in elevation	Stiffness discont. in elevation	Struc discont. in elevation	Struc discont. in elevation (fire)	Struc discont. in elevation

(continued)

Table 4.9 (continued)

	School	Two Apartment buildings	Hotel Eliki	Apartment Building	Apartment Buildings	High-rise Apartments	Student Dormitory	CTV	PGC
Collapse typology	Col-beam joint failure	Disintegration	Collapse of one wing Column Fail. (pancake)	Partial collapse; Column fail. & overturning	Column Fail. (pancake)	Partial collapse; Column fail. & overturning	Column Fail. (pancake)	Column fail. (pancake)	Column fail. (pancake)
Reported Occupants	302	40	157 (38)	24	2652			158	38
Killed Occupants	285	19	10	16	2035			119	16
Fatality rate (%)	**94**	**48**	**6** (26)	**66**	**77**			**75**	**42**
References	Noji ('90)	Noji ('90)	Pomonis (unpublished–GEMECD)	Pomonis (unpublished–GEMECD)	Johnson ('98)		EEFIT (2009)	Pomonis and others ('11)	Pomonis and others ('11)

4.3.3 Extreme Cases of RC Collapses

Over the last 40 years, there have been some individual catastrophic collapses which have been main drivers of the final fatality numbers in particular events, for example, in the 1988 Armenia earthquake, the 1995 Neftegorsk and the 2011 Christchurch earthquakes. Although difficult to apply in a loss estimation model, since these collapses are usually attributed to specific poor construction and code implementation, failures like the 72 apartment buildings of mixed precast concrete frame-panel construction in Leninakan, Armenia, issue a stark warning on rapid construction and correlation of buildings worldwide.

On February 2, 2004, an 11-storey RC frame apartment building collapsed in a pancake manner under its own weight in Konya, Turkey. Figure 4.15 shows a photo of the collapsed building. The volume loss is estimated at 80 %. In this collapse, 92 out of a total of 121 persons who were inside the building lost their lives, and 29 casualties were rescued from the rubble, nine hospitalised patients had crush syndrome i.e., a fatality rate of 76 % (Altintepe et al. 2007).

More recently, an assessment of the two catastrophic collapses after the February 2011 Christchurch earthquake was carried out by Pomonis et al. (2011). Table 4.10 shows a comparison of the findings on injuries and fatalities from this event using the estimates from USGS PAGER system and the original University of Cambridge models (Coburn et al. 1992). In particular, it shows the differences between fatality rates which have been derived from a selection of varying collapse mechanisms and those from a single catastrophic collapse. This study has among other things shows that the mode of collapse, does affect the extent of volume loss, the fatality rate and injury severity distribution among the survivors.

Fig. 4.15 General view of the totally collapsed Zumrut Apartment Building in Konya City, a non-earthquake collapse (*Source* Özdemir 2008)

Table 4.10 Casualty distribution within the collapses for two mid-rise RC frame buildings during the Christchurch earthquake of February 2011 (Pomonis et al. 2011)

	UI	I1	I2	I3	I4	I5
	No injury	Light injury	Moderate injury	Serious injury	Critical injury	Killed
PGC	7 (18.9 %)	5 (13.2 %)	2 (5.3 %)	2 (5.3 %)	3 (7.9 %)	16 (42.1 %)
CTV	17 (10.8 %)	1 (0.6 %)	11 (7.0 %)	3 (1.9 %)	2 (1.3 %)	119 (75.3 %)
SUM[a]	45 (12.2 %)	0 (3.1 %)	13 (6.6 %)	5 (2.6 %)	5 (2.6 %)	131 (**68.9 %**)
PAGER	9.7 %	30.0 %	27.0 %	5.0 %	0.3 %	28.0 %
Difference	**−15.9 %**	**30.0 %**	**27.0 %**	**3.9 %**	**−0.3 %**	**−46.4 %**
UCAM 1992	17.1 %		23.4 %		2.1 %	73.0 %

[a]The total does not add up to 100 % as there few cases with unknown severity of injuries (the deaths are all accounted for)

Table 4.11 Fatality rate assignments for low to medium rise concrete frames

Reinforced Concrete	Typical volume loss (%)	Fatality rates (% of occupants)	Earthquake source
No code low-rise	>10	10–15	Italy 1970–2000s
No code mid-rise	>20	15–30	Kocaeli 1999
No code high-rise	>50	40–60	Romania 1977
Low code low-rise	>10	5–15	Wenchuan 2008, Kocaeli 1999
Low code mid-rise	>20	10–25	Kocaeli 1999, Aeghion 1995
Low code high-rise	>50	30–50	Mexico 1985, Bucharest 1977
Shear wall[a] low-rise	10–20	5–15	Chi Chi 1999
Shear wall mid-rise	<20	10–20	Chi Chi 1999
Shear wall high-rise	<20	10–25	Chile 2010, Bucharest 1977
Catastrophic collapses	***>60***		
Low-rise		***40–60***	Kocaeli 1999
Mid-rise		***50–70***	Christchurch 2011
High-rise		***60–80***	Sakhalin 1995

[a]Low confidence estimates for shear wall structures from very limited data

As shown and as expected, the median value models such as the PAGER semi-empirical model do not capture the high lethality rate of occupants in such catastrophic collapses, however, in future loss estimation models, perhaps some assessment of this low-probability, high-consequence statistic could be incorporated as a bi-modal distribution, with the introduction of damage categories and distributions of fatalities in the context of volumetric reductions of failed buildings.

Based on the empirical evidence presented in this section, the fatality rates have been assigned for the following reinforced concrete frame buildings (Table 4.11).

4.4 Steel

There are very few data for assessing casualties from the collapse of steel frame buildings. As reported in the ShakeOut report (Jones et al. 2008), significant damage to steel frame buildings has only been identified in three earthquakes: the 1985 Mexico City earthquake, the 1994 Northridge earthquake and the 1995 Kobe earthquake. Only the Mexico City earthquake resulted in the collapse of a steel building (the 21-storey Pino Suarez building) and significant casualties, but there is no documentation available to systematically account for those observations. More recently we have seen the complete collapse with very high volume loss of a steel frame industrial warehouse building in Northern Italy (May 20, 2012 Emilia Romagna earthquake) which caused the death of two guards that were inside the building during the night time.

Although no fatality data are available from these specific buildings, references on the structural analyses of the 21-storey Pino Suarez building collapse in Mexico City (Ger et al. 1993) and steel building failures in Kobe, Japan have been examined to infer possible volumetric reductions and therefore fatalities from steel frame building collapses (Fig. 4.16). In addition, analytical studies carried out at Caltech in California have been consulted (Krishnan and Muto 2012).

The failures of most low to mid-rise steel buildings in Kobe were attributed to poor design practice. Systematic failures at weld connections and in the actual steels elements near welds after yielding were observed (Youssef et al. 1995). Brittle fractures were also found in large size steel members, some with cracks of nearly 15 mm. This damage was observed in some high-rise apartment buildings in a large residential development after the Kobe earthquake of 1995. Japanese engineers concluded that the behaviour of thick steel members is significantly affected by size and the small-scaled models tested in the laboratory did not reflect its true behaviour.

Steel moment frame buildings built before 1994 were found to form cracks in their connections during the 1994 Northridge earthquake and are now commonly known as pre-1994 welded-steel moment-frame buildings. These failures have led engineers to believe that a complete collapse as shown in Fig. 4.17 is possible where plastic hinges form in steel buildings (Muto and Krishnan 2011).

Following this argument, in the Los Angeles ShakeOut scenario, high-rise steel frame building collapse-related casualty estimates were generated from a model

Fig. 4.16 *Top* Collapse of the 21-storey Pino Suarez building on top of an adjacent 14-storey building after the Mexico earthquake of 1985 (Credit: EERI); *bottom* the mid-storey collapse of the SRC building (it is the old Kobe Town Hall in Chuo-ku) due to brittle weld connections after the Kobe earthquake in 1995. (*Source* Charles Scawthorn)

derived for complete collapses of 5–10 storey non-ductile concrete buildings from Turkey (Seligson et al. 2006). However, this is a very conservative estimate. As if these complete collapses are possible, as postulated by Muto and Krishnan, the fatality rate of these buildings will be much more than the assumed 8 % in the

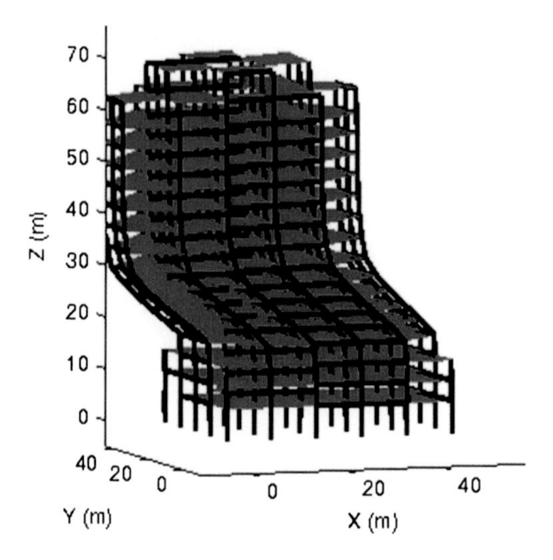

Fig. 4.17 Typical mechanism of collapse from the simulation of the existing building subjected to strong ground motion (synthetic 3-component motion at Northridge from an 1857-like Mw = 7.9 earthquake on the San Andreas Fault). *Deformations are scaled by a factor of 5 for visual clarity* (*Source* Muto and Krishnan 2011)

ShakeOut model, especially with the added complexity of the search and rescue task of very high occupancy buildings and difficult access.

The challenge to future casualty research is to realistically portray the lethality potential from the failures of 'untested' tall steel frame buildings, perhaps based on engineering judgment. Two crucial questions need to be addressed: what is the likelihood of the particular ground motion that would excite the brittle failure mechanism described in Muto and Krishnan (2011) and what is the likelihood of this mode of pancake collapse? Are catastrophic collapses witnessed in New York City after the World Trade Center attacks realistic? These hypotheses have real and potential catastrophic implications for high occupancy office buildings. As shown in the assessment of pancaked reinforced concrete buildings, the fatality rates can be as high as 80 % and steel being more ductile could collapse under its own weight leaving even less voids for its occupants.

4.4.1 Light Metal Frame

There is very little fatality data available for light metal frame buildings but 86 did collapse after the Kobe earthquake of 1995; some may have died in these buildings but they tended to be small commercial buildings and most were empty at the time of the earthquake at 5:46 am.

Due to a lack of data, the suggested fatality rates shown in Table 5.1 for steel buildings are based on observations of reinforced concrete frames with similar volumetric reduction due to collapse.

References

Altintepe L, Guney İ, Tonbul Z, Türk S, Mazi M, Ağca E, Yeksan M (2007) Early and intensive fluid replacement prevents acute renal failure in the crush cases associated with spontaneous collapse of an apartment in Konya. Ren Fail 29(6):737–741

Berg GV, Husid RL (1971) Structural effects of the Peru earthquake. Bull Seismol Soc Am 61 (3):613–631

Brzev S, Astroza M, Yadlin MOM (2010) Performance of confined masonry buildings in the February 27, 2010 Chile Earthquake: EERI, at www.confinedmasonry.org

Campobasso CP, Falamingo R, Vinci F (2003) Investigation of Italy's deadliest building collapse: forensic aspects of a mass disaster. J Forensic Sci 48(3):635–639

Cardoso R, Lopes M, Bento R, D'Ayala D (2011) Historic, braced frame timber buildings with masonry infill ('Pombalino' buildings) in World Housing Encyclopaedia, EERI and IAEE, Report no 92

Coburn AW, Spence RJS, Pomonis A (1992) Factors determining casualty levels in earthquakes: mortality prediction in building collapse. In: 10th Proceedings of the world conference of earthquake engineering, Madrid, Spain. 19–25 July 1992

De Bruycker M, Greco D, Lechat MF, Annino I, De Ruggiero N, Triassi M (1985) The 1980 earthquake in Southern Italy: morbidity and mortality. Int J Epidemiol 14(1):113–117

Durkin ME, Murakami H (1988) Casualties, survival and entrapment in heavily damaged buildings. In: Proceedings of the 9th world conference on earthquake engineering, Tokyo, Japan, 2–9 August 1988, Vol. 8, pp. 977–982

EEFIT (Earthquake Engineering Field Investigation Team) (2009) The L'Aquila, Italy earthquake of 6 April 2009: a field report by EEFIT: earthquake engineering field investigation team. Institution of Structural Engineers, London

EEFIT (Earthquake Engineering Field Investigation Team) (2010) The Haiti earthquake of 12 January 2010: a field report by EEFIT: earthquake engineering field investigation team, Institution of Structural Engineers, London

FDMA (2008) (Fire and Disaster Management Agency) Japan

Georgescu ES, Pomonis A (2011) Emergency management in Vrancea (Romania) earthquakes of 1940 and 1977; Casualty patterns versus search and rescue needs: proceedings of the international emergency management society 2011 Annual Conference, Bucharest, Romania, June 2011

Ger JF, Cheng FY, Lu LW (1993) Collapse behavior of Pino Suarez building during 1985 Mexico city earthquake. J Struct Eng 119(3):852–870

Jones LM, Bernknopf R, Cox D, Goltz J, Hudnut K, Mileti D, Perry S, Ponti D, Porter K, Reichle M, Seligson H, Shoaf K, Treiman J, Wein A (2008) The ShakeOut scenario, U.S. Geological Survey Open-File Report 2008-1150 and California Geological Survey Preliminary Report 25, at http://pubs.usgs.gov/of/2008/1150/

Krishnan S, Muto M (2012) Mechanism of collapse of tall steel moment frame buildings under earthquake excitation. J Struct Eng 138(11):1361–1387

Kuwata Y, Takada S, Bastami M (2005) Building damage and human casualties during the Bam-Iran earthquake. Asian J Civ Eng (Build Hous) 6(1–2):1–19

Langenbach R (2000) Intution from the past; What we can learn from traditional construction in seismic areas at http://www.conservationtech.com/IstanCon/keynote.htm

Langenbach R (2009) Don't tear it down! preserving the earthquake resistant vernacular architecture of Kashmir. Oinfroin Media. ISBN-10: 9780979680717

Muto M, Krishnan S (2011) Hope for the best. Prepare for the worst: response of tall steel buildings to the ShakeOut scenario earthquake: ShakeOut Special Issue, Earthquake Spectra, vol 27, no 2, pp 375–398

NIBS-FEMA (National Institute of Building Sciences-Federal Emergency Management Agency) (2006) HAZUS-MH MR2 Technical Manual

Okada S (1996) Description of indoor space damage degree of building in earthquake. In: 11th Proceedings of the world conference in earthquake engineering, Acapulco, Mexico, 1996, Acapulco, Mexico. CDROM, paper no 1760

Özdemir A (2008) A geological and geotechnical investigation of the settlement area of Zümrüt Building (Konya, Turkey) which caused 92 fatalities due to its collapse. Environ Geol (2008) 53:1695–1710

Petal MA (2004) Urban disaster mitigation and preparedness: the 1999 Kocaeli earthquake: Los Angeles, University of California, Department of Urban Planning, PhD dissertation

Plafker G, Ericksen GE, Concha JF (1970) Geological aspects of the May 31 Perú earthquake. Bull Seismol Soc Am 61(3):543–578

Pomonis A, Sakai S, Coburn AW, Spence RJS (1991) Assessing human casualties caused by building collapse in earthquakes. In: Proceedings of the international conference on the impact of natural disasters, UCLA, Los Angeles, USA, July 1991

Pomonis A, So E, Cousins J (2011) Assessment of fatalities from the Christchurch New Zealand earthquake of February 22nd 2011: seismological society of America 2011 annual meeting, Memphis, Tennessee, USA, 13–15 April 2011

Rota M, Penna A, Strobbia CL (2008) Processing Italian damage data to derive typological fragility curves. Soil Dyn Earthq Eng 28(10–11):933–947

Seligson HA, Shoaf KI, Kano M (2006) Development of casualty models for non-ductile concrete frame structures for use in PEER's performance-based earthquake engineering framework. In: 8th proceedings of the 100th anniversary earthquake conference commemorating the 1906 San Francisco earthquake and U.S. national conference on earthquake engineering, San Francisco, California 18–22 April 2006

So EKM (2009) The assessment of casualties for earthquake loss estimation. Cambridge, University of Cambridge, PhD dissertation

Spence R, So E, Jenny S, Castella H, Ewald M, Booth E (2008) The Global Earthquake Vulnerability Estimation System (GEVES): an approach for earthquake risk assessment for insurance applications. Bull Earthq Eng 6(3):463–483

Suganuma K (2006) Recent trends in earthquake disaster management. Q Rev Sci Technol Foresight Cent 19:91–106

Sun BT, Zhang GX (2010) The Wenchuan earthquake creation of a rich database of building performance. Sci China Technol Sci 53(10):2668–2680

Yadlin MO (2011) University of Chile

Youssef NFG, Bonowitz D, Gross JL (1995) A survey of steel moment- resisting frame buildings affected by the 1994 Northridge earthquake: NIST report no NISTIR 5625, National Institute of Standards and Technology, United States Department of Commerce Technology Administration, Washington

Chapter 5
Conclusions

A set of judgment-based fatality rates in collapsed buildings for implementation in earthquake loss estimation models have been compiled and shown in Table 5.1. The fatality ranges are presented as a percentage of the total number of occupants of a completely collapsed building (volume loss of 10 % or greater). Some of these rates were based on unique observations from particular structural collapses, whereas others are based on averaging over a period of time and several events. The table shows the building types that have been evaluated based on empirical data to date and the building types have been mapped with the PAGER STR taxonomy (Jaiswal and Wald 2008). Furthermore, for USGS PAGER's semi empirical model, where fatalities are assessed for specific classes of buildings and not individual buildings, a suggested fatality rate has been proposed in the right hand column of Table 5.1.

This work is still on-going and it is hoped that a search for local unpublished data through regional programs of the Global Earthquake Model (GEM) initiative will verify existing rates and add to structural types where there is currently little information. Future work would help improve these mean fatality ranges, especially in narrowing large ranges.

5.1 The Way Forward

The fatality rates ranges presented in Table 5.1 can be directly applied to earthquake loss estimation models like the USGS PAGER's semi-empirical model, where the main contribution to estimates of deaths is assumed to be the collapse of buildings due to ground shaking. Yet, there are obvious limitations to the numbers presented. Although compiled based on a thorough review of historical evidence, they are still based on opinions of the reviewer of available data. Given the interest in learning from global earthquakes amongst seismologists, engineers and social scientists, opportunities exist to capture useful casualty data for the future to supplement and inform the findings from this book. It is therefore important to devise a methodology of assessing new data and a way to use this information to update existing assumptions on lethality potential of particular collapsed building types. In particular, and the next phase of work would include matching fatality rates to an index

E. So, *Estimating Fatality Rates for Earthquake Loss Models*,
SpringerBriefs in Earth Sciences, DOI 10.1007/978-3-319-26838-5_5

Table 5.1 Judgment-based fatality rates for use in loss estimation models

	Building Classes	PAGERSTR	Typical volume loss (%)	Fatality ranges from literature (%)	Suggested Fatality Rate (for PAGER) (%)
	Light Timber	W1			
1	With light roof		<10	0.25–1	0.5
2	With heavy roof		>50	0.75–3	1.0
	Heavy Timber	W3			
3	With light roof		<10	0.5–1	0.5
4	With heavy roof		>50	2–3	2.0
	Weak Masonry				
5	Adobe light roof	A5	<50	5–15	5.0
6	Adobe heavy roof	A4	>75	20–90	65.0
7	Irregular stone with wooden pitched roofs (low-rise)	RS2	40–60	5–20	10.0
8	Irregular stone low-rise concrete slab roofs	RS5	>70	10–40	30.0
	Load-bearing masonry				
9	European	DS2	>30	3–12	5.0
10	Asia	DS4	>50	10–25	15.0
	Reinforced masonry				
11		RM2L	>10	2–8	4.0
12		RM2 M	>20	15–40	25.0
13	Confined masonry		<10	2–6	3.0
14		Extreme collapse (pancake)	>60	10–40	
15	Mixed		>30	15–20	
	Concrete frame				
16	No code, low-rise	C4L	>10	10–15	15.0
17	No code, mid-rise	C4 M	>20	15–30	30.0
18	No code, high-rise	C4H	>50	40–60	50.0
19	Low code, low-rise	C3L	>10	5–15	12.0
20	Mid-rise	C3 M	>20	10–25	25.0
21	High-rise	C3H	>50	30–50	40.0
	Concrete shear wall				
22	Low-rise	C2L	10–20	5–15	5.0
23	Mid-rise	C2 M	<20	10–20	10.0
24	High-rise	C2H	<20	10–25	15.0

(continued)

Table 5.1 (continued)

	Building Classes	PAGERSTR	Typical volume loss (%)	Fatality ranges from literature (%)	Suggested Fatality Rate (for PAGER) (%)
	Catastrophic collapses	*Low-rise*	*>60*	40–60	40.0
		Mid-rise	*>60*	50–70	50.0
		High-rise	*>60*	60–80	70.0
	Steel frames				
25	Low-rise	S1L	<10	10–15	5.0
26	Mid-rise	S1 M	>20	15–25	8.0
27	High-rise	S1H	>20	20–25	25.0
28	Braced frames	*Catastrophic collapse*	>60	50–70	40.0
29	Light metal frame	S3	<10	4–6	10.0

of volumetric losses of different building types currently being developed, in conjunction with a casualty-inducing damage scale.

For three of the 25 assessed earthquakes, casualty surveys were carried out by the author amongst surviving households of the earthquakes in the affected area to gather information on the causes of injuries and deaths. The data collected from these surveys are extremely rare as for a sample of the affected population; the surveys have been designed to explore the causal pathways of injuries and deaths, linking the damage to the buildings they are housed in or the local setting at the time of the event to their post-event circumstances. All of these surveys also include non-damaged buildings and affected people who are not injured or killed.

5.1.1 GEM Earthquake Consequences Database (GEMECD)

A consortium of 10 international partners, led by the author have worked on a three-year project to produce the GEM earthquake consequences database (GEMECD). The aim was to assemble and store in a structured and web-accessible way both data (including photographs) already acquired and data yet to be acquired following future events that show the consequences of earthquakes, including building damage, damage to lifelines and other infrastructure, ground failure, human casualties, social disruption, and financial and economic impacts. The GEMECD project started in November 2010 and houses consequences information for 71 global events. The main objective of the GEMECD is to capture important lessons in the form of standardised data and formats to inform future loss estimation models and design documents.

The study of fatality rates from empirical observations of past events is ongoing and it is hoped with the rigorous data collection for GEMECD, previously unpublished data on fatalities from building collapses can be attained and included in the future as part of GEM. There is a set of guidelines for future data collection including standardised and minimum requirements for collecting data on casualties. It is hoped that these consistent datasets will address the current situation of a lack of good quality casualty data and the methodology described in this book on informing expert opinion information with empirical distributions of fatalities in collapsed buildings can be incorporated into future probability loss estimation models.

5.2 In Closing

Earthquake loss estimation models provide a means to present current and future risk to decision makers and must therefore include realistic estimates of physical and social losses. The judgment based fatality rates proposed in this book have been derived from real event data in the past 40 years and are intended to help gauge potential fatalities in future earthquake scenarios and building inventories. Poor construction, like diseases, can be eradicated. The scientific community should strive to eliminate preventable deaths in earthquakes through science, education and construction. Whilst the hazard itself is not preventable, the challenge is how as scientists we use our skills to influence decisions made by people going about their everyday lives and create a culture of seismically resistant housing.

Reference

Jaiswal KS, Wald DJ (2008) Creating a global residential building inventory for earthquake loss assessment and risk management: US geological survey open-file report 2008-1160, 113 p